BASIC MASONRY
ILLUSTRATED

BY THE EDITORS OF
SUNSET BOOKS AND
SUNSET MAGAZINE

**LANE PUBLISHING CO.,
MENLO PARK, CALIFORNIA**

Masonry colors and textures make for harmony in the garden. Here, low brick retaining wall, brick-in-sand paving, and rugged stone wall complement a lush, romantic tumble of flowers.

**EDITED BY
SCOTT FITZGERRELL**

Design: Cynthia Hanson

Illustrations: Rik Olson

Cover: Photographed by Jack McDowell. Cover design by Zan Fox.

Editor, Sunset Books: David E. Clark

First Printing May 1981

Special thanks . . .

. . . to landscape architect Bill Kapranos for generously sharing his knowledge of structure and design with us; and the following individuals and companies for their assistance in the preparation of this book: Holly Lyman Antolini; Andy Clovich, Port Costa Brick; Wallace Evans, San Jose Brick; Tom Harley, Hans Sumpf Co.; Donnan Jeffers; C. W. Kraft, Kraftile; Joe Morey, Peninsula Building Materials; Muller Supply Co.; Bill Otey; Paige Structural Glass Co.; The Tile Shop; Western Sand & Brick.

. . . and a special thank-you to Hilary Hannon for her work in assembling the color section.

Concrete block lies under the paint of this simplest of walls. Blocks, grouted with concrete, make an effective barrier against the noise of an adjoining busy street. Landscape architect: Peter Lockhart.

Exposed-aggregate concrete makes a practical, beautiful surface. Here, effect is enhanced by stone planters and step edgings. Landscape architect: Mary Gordon.

Brick pool deck and low, brick-edged planting bed combine harmoniously with old garden wall. Brick coping is mortared, providing a secure edge for swimmers on one side and for brick-in-sand paving on the other. Landscape architect: Mary Gordon.

Cool patio garden gets its character from masonry. Tying everything together is brick — used for wall, entry paving, and concrete slab dividers. Landscape architect: W. David Poot.

CONTENTS

THE MANY FACES

Brick-in-sand Paving, p. 60

Poured-concrete Walls, p. 67

Dry Stone Walls, p. 52

Adobe Walls, p. 48

Mortared Stone Walls, p. 52

For the homeowner, building with masonry — brick, block, stone, tile, and concrete — has unique advantages. No other group of materials offers masonry's combination of beauty, utility, ease of maintenance, and durability — and the techniques necessary to build with these materials are few and fairly easy to master.

Masonry is permanent. It is highly resistant to natural deterioration by wind, water, fire, sun, and pests. Indeed, man's rough hand has proved the undoing of more masonry structures than has the relatively gentle hand of nature. Most surviving structures of the ancient world are of masonry, and what deterioration they show is due more to warfare — and succeeding generations' penchant for a handy supply of building materials — than to their battles against the elements.

Masonry materials are natural materials, and this is the key to their appeal. Earth, air, fire, and water combine to make bricks, blocks,

OF MASONRY

Concrete Pavers, p. 57

Brick Walls, p. 36

Tile in Mortar, p. 63

Poured-concrete Paving, p. 74

Mortared Brick Paving, p. 62

Concrete Block Walls, p. 44

concrete, even glass and stone. Whether the forming process takes place in the earth or in a kiln, it is essentially the same, and the products have a truly elemental appeal.

This book leads you step-by-step through the world of masonry. Starting on the next page, there is a color section, a "catalog" of masonry

building ideas and materials. Three how-to chapters follow, two on unit masonry—brick, concrete block, adobe, stone, and tile—and one on poured concrete. In them, you'll find step-by-step instructions for building walls and pavings. The photos above are your visual guide to these chapters. A chapter on mainte-

nance and repair of masonry comes next, followed by a project section designed to give you ideas for applying your new knowledge.

Masonry work is work well rewarded. Properly built, your finished project can last as long as the earth from which it is made.

Let's meet the materials.

BRICK

Made of various clay mixtures, bricks once were molded by hand but now are usually extruded: the clay is forced through a die, then cut to size with wires. After drying, the bricks are fired, or "burned," in a kiln. Once they've been fired they become permanent; they can no longer be reduced to soft clay.

In cost, bricks are moderate to expensive. They're easily handled, and they range from rustic to elegant in their effect. In addition, sizes range widely. The photo above shows only a few of them. Since it is uneconomical to transport bricks very far they are usually locally made, and types and sizes tend to vary from one region to another.

Brick walls must be mortared, and they must be built on concrete foundations. Building codes, especially in earthquake-prone areas, require steel reinforcement in walls above a specified height.

A properly engineered brick wall needs little care; its life is measured in generations.

1 **Wire-cut common,** in economical paver thickness.

2 **Smooth-face common,** for all kinds of general building.

3 **Used brick,** best for appearance only, since strength is unknown.

4 **Smooth-face "soap,"** half as wide as a standard brick.

5 **Simulated used,** or "rustic," splashed with mortar and stain.

6 **Split paver, speckled,** half as thick as a standard brick.

7 **Wire-cut Norman,** chocolate color, longer than a standard brick.

8 **Firebrick,** for use in fireplace linings, other areas of high heat.

9 **Smooth common,** chocolate color for a subdued effect.

10 **Smooth common** in another color variation—fawn.

11 **Cored jumbo,** for rapid building with less effort, better mortar bond.

12 **Cored standard,** buff color, wire-cut surfaces.

13 **Cored Norman,** combining advantages of cores with a low silhouette.

14 **Flashed common,** discolored by flame contact in the kiln.

Handsome combination of headers, soldiers, and stretchers, together with carefully laid grille, gives common-bond wall an uncommon look. Decorative pilasters reinforce the wall at opening and corner. Mortared borders retain the brick-in-sand walk. Architects: Fisher-Friedman.

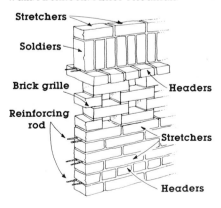

Stretchers
Soldiers
Brick grille
Headers
Reinforcing rod
Stretchers
Headers

White-painted brick wall, continuing into a low planter, makes a pleasing color contrast to mortared basketweave brick entry walk and exposed-aggregate pool deck. Landscape architects: Royston, Hanamoto, Alley & Abbey.

Brick . . . for beautiful walls

Curved garden wall uses flashed bricks with raked joints for out-of-the-ordinary color and texture. "Pop-outs," pilasters, and header cap all lend visual variety to the long facade. Drawing details the construction. Designer: Thomas Ceranic.

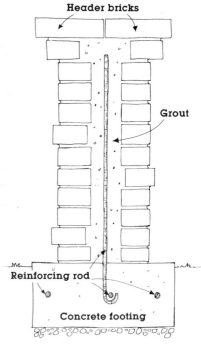

Old garden wall, with its arch and graceful steps, may show signs of wear and tear, but it's still full of ideas for the modern mason. Beautiful Flemish bond pattern, with pilasters and rowlock cap, adds strength. Arch was built over a wooden form. Integral retaining wall braces wall from the back and acts as a border for the steps.

Dignified transition from sidewalk to house is provided by low retaining wall that also prevents runoff from lawn sprinklers. Designer: Louis Marano.

Curved retaining wall of common brick becomes angular to create an alcove for a bench. Wall is laid in running bond, 8 inches thick, with a simple cap of headers. Landscape architect: William Louis Kapranos.

Paving with character

Basketweave paving has held up for years. Bricks were laid in sand with mortared edgings. Handsome pattern has true interwoven look, requires cutting at edges.

Bricks recycled from a patio torn up during remodeling get a new lease on life in attractive driveway. Timeworn bricks are laid in mortar over a concrete slab.

Bricklayer's delight, this three-level entry court ably shows off the versatility and attractiveness of brick. Mortared paving is laid in a herringbone pattern with railroad tie steps and edgings. Low walls are built in running bond with raked joints. Designers: Galper/Baldon.

Simple and effective, this patio in red common brick was easy to build. Bricks are laid in jack-on-jack bond in a sand bed with redwood edgings. Sand swept into the joints keeps paving tight. Designer: Roy Rydell.

Attractive outdoor dining space in chocolate brick is the heart of an elegantly simple garden. Bricks are laid in sand in a herringbone pattern with mortared edgings that extend around flower beds. Designer: Louis Marano.

Gently curving walkway in used brick leads invitingly to sunny patio at top. Steps are brick-in-sand with mortared risers; low side walls hold everything in place. Landscape architect: William Louis Kapranos

Brick ideas, from the ground up

Circular firepit provides cheerful focal point for poolside parties and extends use of area well into evening. Outer circle is expertly cut and fitted, as is mortared pool deck. Simple stack of unmortared bricks retains stones covering a gas burner. Landscape architect: Gene Kunit.

Fountain adds classical style and music of splashing water to a small garden. Owners had lion's head cast from an antique mold; they tiled fountain's interior themselves. Flashed brick with raked joints is used throughout. Designer: Mary Gordon.

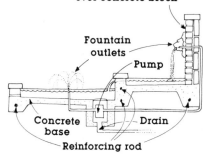

Brick veneer over concrete block

Fountain outlets

Pump

Concrete base

Drain

Reinforcing rod

Intimate alcove just off entry to house gets its charm from mossy used-brick paving laid in basketweave pattern, with low brick planters and seat supports. Designer: Roy Rydell.

Barbeque's straightforward design includes storage space behind doors to left of drop-in cooking unit. Tile counter provides ample area for food preparation. Bricks have a simulated used finish and are laid in dark mortar with raked joints. Designer: Janet Pollock.

Used-brick veneer sets off a small arched fireplace. Herringbone inset above arch contrasts with sturdy pilasters laid in running bond. Designer: Bob Dutton.

Two for the expert. Chimney at left harks back to the days of fancy brickwork; design required a careful hand and much cutting. Architects: Fisher-Friedman. Low steps, curving planters, and elaborate paving (above) merge into a sophisticated and distinctive entry. Architects: Sandy & Babcock.

BLOCK

When a masonry unit isn't a brick, it's a block. Building and paving blocks of concrete, adobe, and glass are the most common.

Concrete blocks come in many forms, and one can be found for nearly any project, from rugged foundation walls to lacy screens.

The large size and relatively light weight of concrete blocks make for rapid building. Their hollow cores, or cells, make reinforcement simple — steel bars are run inside and secured with mortar or grout.

For most purposes, concrete block is less expensive than brick. If its appearance doesn't suit, you can veneer it with brick, stone, or other material. It can also be sandblasted to reveal its aggregate, or it can be painted, stuccoed, or plastered.

Interlocking concrete pavers, developed in Europe, are increasingly popular in this country. They make a rugged, low-maintenance surface that is easy to lay. Turf blocks give you the option of adding greenery to the surface; they combine traffic-carrying capability with the look of a lawn. Cost is about the same as for bricks.

Adobe, the mud brick of the Southwest, is one of the world's oldest building materials. Modern adobe blocks are stabilized with asphalt emulsions to make them impervious to water. In the garden, adobe adds a character and mellowness unmatched by other masonry materials.

Adobe is inexpensive unless it must be shipped long distances; then, costs may be high. One solution is to make the blocks yourself — it's laborious, and codes may restrict you, but the Mesopotamians did it, and so can you.

Glass blocks lie on the opposite end of the masonry spectrum from adobe; they are the least earthy of masonry materials. Popular during the '20s, '30s, and '40s, they are currently staging a comeback.

The blocks are made of two hollow sections fused together with a partial vacuum inside. This accounts for their good insulating and acoustical qualities. Mortar-bearing surfaces are treated to provide a good bond. Installation can be tricky; to do it consult a pro.

1 **Standard block** is 8 inches wide, comes in fractional units shown.

2 **Slump blocks** come tinted and in regular gray concrete.

3 **Screen blocks** admit light and air.

4 **Other widths** shown are standard and "corner return" units in 6-inch width, and fractional units in 12-inch width.

5 **Turf blocks** allow both traffic and planting, are easy to lay.

6 **Interlocking concrete pavers** make long-wearing surfaces.

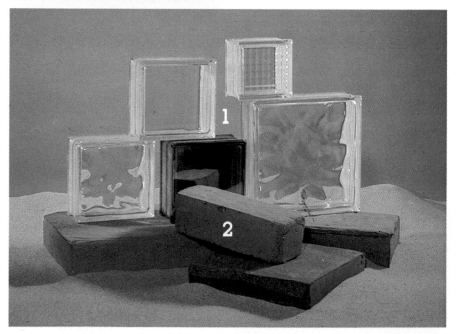

1 **Glass block** is the least earthy of masonry materials. Shown are wavy, clear, cross-hatch, and solar-reflective types in 6, 8, and 12-inch sizes.

2 **Adobe** is the earthiest material. Shown, clockwise from left, are 12-inch wall unit, 4-inch veneer unit, 8-inch wall unit, and 12-inch paver.

Block walls . . . earthy to urbane

Covered with stucco, concrete blocks combine to form a unified design in this curving garden wall. Screen blocks near the top admit light and air. Drawing shows how it was done. Designer: Siegfried Bartusch.

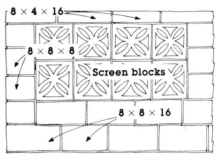

Soft earth tones of this wall of adobe veneer are picked up by the tile floor. Seat over wood storage space extends into the fireplace and is supported by a series of iron rods.

Glass block was the perfect solution when an exterior wall became an interior one during remodeling. Wall allows light (from window and skylight in new hallway) into existing bathroom, yet provides needed privacy. Architect: Daniel Solomon.

A potpourri of block pavings

Decorative and durable, interlocking concrete pavers are well adapted to high load applications, such as driveways.

Patio of concrete pavers remains stable, thanks to redwood edgings and dividers. Edgings are anchored to underlying concrete blocks; dividers are bolted to small concrete footings.

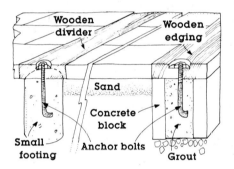

Adobe's natural warm color contrasts with splashy display of multicolored impatiens along this winding garden path. Adobe blocks in curved section were easy to cut and were laid on a bed of crushed rock.

The adaptable concrete block

Looking like stone, split-face concrete blocks form retaining wall for bark-covered play area. Landscape architect: Ted Sutton.

Decorative screen wall has raked joints for shadow effect. Dark mortar shows off stone veneer. Wall on right is made of slump blocks.

Dramatically topped with a large flat stone, octagonal tile-lined fountain was built with half-height concrete blocks. Fountain bowl is cast concrete. Garden wall echoes fountain's construction. Landscape architect: Peter Lockhart.

Adobe-like wall of white-painted slump blocks winds around spa. Four-inch openings were created with half-size blocks. Expanse of exposed-aggregate paving is relieved by decorative strips of brick. Landscape architects: Peterson-Jones.

STONE

Nature's own building material is rarely used structurally now, because of its expense. But where rugged good looks are important, there is nothing like stone. When you build with stone, you build with nature.

Stoneyards supply uncut (rubble) or cut (ashlar) stone of various types. The latter is the more expensive.

Synthetic stone veneers, made of cement mixtures, are often surprisingly like the real thing. You may want to consider them for their cost and labor-saving qualities. Because of their light weight, there usually is no need to reinforce the structure they will cover.

1 **Rubble stone,** here ranging from rough to smoothly worn granite "baldies," is used with little trimming.

2 **Stucco veneer** gives a remarkable impersonation of the stone just to its right; it's useful where weight and cost are factors.

3 **Roughly squared** granite cobblestones serve well for both walls and paving.

4 **Ashlar** sandstone can be laid like brick, makes a durable veneer.

5 **Imitation sandstone** resembles the real thing, costs much less, comes in both flat and corner shapes.

6 **Slate** can range widely in color; is often used for interior floors.

7 **Flagstone,** produced from the same rock as the ashlar veneer above, makes elegant outdoor paving.

8 **Fieldstone** includes a wide variety of sizes and types suitable for both walls and paving.

Rough stone wall, lush garden, moody sea— all conspire for a perfect setting. The natural harmony of sea-worn beach stones and rugged coastal landscape was keenly felt by designer-builder Robinson Jeffers, the noted poet. Working alone, he built the sturdy wall and the massive tower it encloses (left) over a period of years.

Stone & water—a natural pairing

Natural river rock veneers a pool deck and retaining wall. Stones, mortared in place over concrete, ease the transition from plantings to pool. Landscape architects: Singer & Hodges.

These stones have been around— Cape Horn, that is. Culled from an old, torn-up street, the stones once ballasted the holds of Victory ships, and before that, square-riggers. Here, they find a resting place in a curving pool wall and planter. Landscape architect: Ted Sutton.

Indoor/outdoor charm of this Japanese bath is enhanced by the use of river rock on tub sides and floor. Architect: Alfred Klyce.

Cobblestone . . . recycling yesteryear

Painstaking placement of irregular cobblestones gives this fireplace its dramatic good looks.
Architect: Doug Dahlin.

Large, roughly squared cobblestones in these gateposts are laid in courses and effectively set off by deeply raked joints. Drawing shows how concrete blocks form the core. Designer: Mary Gordon.

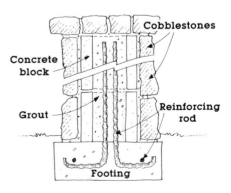

Stone walls . . . casual or formal

Dry-laid rubble wall blooms with a profusion of colorful plants rooted in its earth-packed joints. It acts as a retaining wall for garden above.

Sloping sandstone wall was expertly laid over 50 years ago. Ashlar work such as this is becoming increasingly rare; a search of older neighborhoods can turn up models and inspiration.

Flagstone — nature's own & a clever imposter

Synthetic stone veneer over concrete-block planter and entry wall extends onto house wall to muffle noise from busy street. Cost was less than half that of natural stone. Carefully fitted ungrouted joints minimize exposed mortar; construction is detailed below. Architect: Thomas Lile.

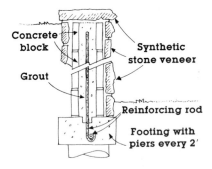

Concrete block

Synthetic stone veneer

Grout

Reinforcing rod

Footing with piers every 2′

Flagstone paving in this handsome circular patio sets off fountain and pool. Fieldstone forms the backdrop for a bas-relief that sends sheets of water into the pool below. Designers: Dick Ammon, Ted Gantz.

TILE

Like brick, tile is a fired-clay product and is available in a great variety of sizes and shapes. Low-fired tile is porous and rather soft compared to high-fired tile, which is more vitreous (glasslike) and durable. The composition of the clay and the heat of the firing make the difference.

Both types can be had either glazed or unglazed. Glaze is a thin, glassy coating bonded to the clay at very high temperatures in the kiln. Glaze adds and intensifies color, and it gives tile texture and durability.

Tile is equally well adapted to pavings and wall coverings. In the kitchen and bath it makes one of the most durable surfaces available for counters, walls, and floors.

1 **Unglazed patio tile** makes durable outdoor surfaces.

2 **Rectangular tile** is available both glazed and unglazed.

3 **Glazed tile,** both decorated and plain, gives a more formal look, is best used indoors.

4 **Handpainted tile,** too soft to sustain traffic, makes decorative accents. Cost is high, but so is beauty.

5 **Tile with "pickets"** lends variety to paving pattern.

6 **Small, colorful tiles** are another way of achieving decorative accents, this time at a cost saving.

7 **Glazed, patterned tiles** make handsome interior floors; wet weather would make them too slippery for outdoor use.

8 **Quarry tile** ranges in color from buff to deep brown; color comes from the clay itself. Tiles are unglazed.

9 **Large rectangular tile** resembles adobe, combines adobe's mellowness with the durability of fired clay.

10 **Flashed quarry tile** is available in hexagonal and rectangular shapes, in addition to the standard square.

Unglazed patio tiles harmonize with brick wall in this enclosed barbecue area. Machine-made tiles are economical and easy to work with. Designer: Bo Tegelvik.

Sand-colored quarry tiles transform plain concrete slabs into an inviting entry. Each square tile is surrounded by "pickets" to create an overall octagonal pattern. Design: Designed Environ, Inc.

Handmade Mexican floor tiles make a beautiful and practical surface. Here, they're accented by kitchen counters of handpainted Mexican tile, and matching window trim. The wine rack is tile, too — unglazed tile drainpipe.
Architect: Alfred T. Gilman.
Designer: Windom Hawkins.

POURED CONCRETE

This is truly the building material of the modern age. From planters to skyscrapers it is all around us. Poured concrete, especially when reinforced with steel, can be made into almost any size and shape.

Concrete is made from cement, sand, aggregate (usually stone and gravel), and water. Once it hardens, it forms a dense, permanent material with enormous compressive strength. The use of steel reinforcement strengthens the material against tension as well, making possible everything from your home's foundation to sweeping freeway ramps and graceful bridges.

For the homeowner, concrete is a logical choice for pavings, garden pools, and small projects. It is literally fundamental to most masonry projects, since it is the material of choice for footings. And its flexibility of both application and appearance makes it a good choice for many other projects as well.

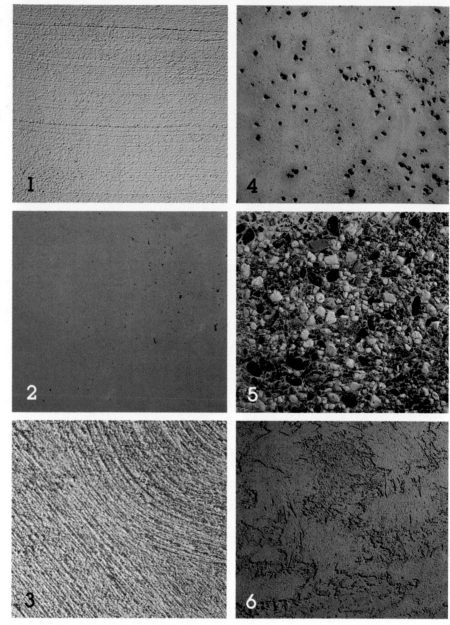

1 **Wooden float** leaves a semismooth texture that is still rough enough to be glare-free and provide traction.

2 **Steel trowel** smooths concrete to a dense, reflective surface suitable for enclosed patios and interiors. This slab was tinted brown.

3 **Broomed surface** is best where maximum traction is needed. Straight and wavy patterns are easy to produce.

4 **Rock salt** left the pockmarks in this sand-colored concrete. Salt is embedded while concrete is soft, later washed out.

5 **Exposed-aggregate** surface is widely used for both safety and attractiveness. Selected aggregate is embedded after concrete is placed; then aggregate is exposed with hose and broom.

6 **Travertine finish** is produced when mortar is dashed onto freshly placed concrete. Troweling the mortar produces a stony, layered look. Here, both concrete and mortar have been tinted to bring out the texture.

Practical paving

Tranquillity of a mountain pool is captured in this spa by an irregular edging of smooth river rock set in seeded-aggregate concrete. Designer: Mary Gordon.

Low concrete wall provides safety, yet doesn't obscure magnificent view. Pool deck is tinted and has salt finish. Landscape architects: Singer & Hodges.

Large concrete steppingstones seeded with smooth aggregate and variegated pebbles make a handsome entry. Spaces between stones invite growth of ground cover, softening edges. Designer: Carolyn Susmann.

Durable steps

Poured-concrete steps march downward through a series of natural rock outcroppings and expertly built stone retaining walls. Simple wood-float finish of steps keeps them from competing with stonework.

Variety in color and texture is provided by poured-concrete steps with railroad tie headers. Ties were used as forms for the concrete; left in place, they make a functional edge. Pressure-treated peeler logs border steps. Landscape architect: William Louis Kapranos.

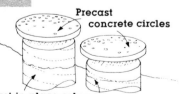

Concrete "mushrooms" flourish in a smart, contemporary entry stairway. Exposed-aggregate slabs were cast separately, then mortared to poured-in-place footings cast in tubular cardboard forms (see drawing). Designer: John Nishizawa.

Precast concrete circles

Cast-in-place columns

Rugged retaining walls

Wooden forms left their mark on poured-concrete wall. Boards were sandblasted to emphasize imprint of wood grain. Extra-thick lip attests to expert casting, as do exposed-aggregate steps. Landscape architects: Wilson and van Deinse.

Transition from driveway to front gate is accomplished with exposed-aggregate steps and with retaining walls that form a series of planting beds. Landscape architect: Robert Babcock.

Creative casting

Straight as a sentry, sturdy gatepost of cast concrete supports gate and anchors wooden fence. Formwork for post is detailed in drawing. Landscape architects: Singer & Hodges.

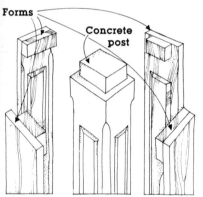

Angular ornamental pool has poured-concrete edge with exposed-aggregate finish. At lower left, house entry slab bridges one end of pool. Architect: Henrik Bull.

Smooth concrete elegantly frames tile-bordered fireplace. Concrete was cast in sectional forms; sand-cement mixture (no large aggregate) ensured a smooth finish. Architect: Don Gentry.

Simple, oriental-style design of fireplace was achieved with three cast-concrete slabs and exposed-aggregate finish. Tile hearth neatly turns corner to become floor of entryway. Architect: Henrik Bull.

Masonry is your best choice when you want a wall that will last — one you can build, then forget.

In this chapter you'll find easy-to-follow, step-by-step instructions on how to build walls of brick, concrete block, adobe, and stone. The instructions apply equally to such projects as planters, borders, steps, hearths, and barbecues — once you've mastered the basic techniques needed to build walls with each material, you'll be able to strike off on your own.

Each material is an example of what is called *unit* masonry; that is, masonry materials that are made in units small enough for one person to handle. (A chapter on *continuous* masonry — poured concrete — begins on page 64.) All unit masonry materials are assembled using mortar.

A discussion of mortar — what it is, how to mix and apply it — starts on the facing page. It is followed by subsections on how to handle brick, concrete block, adobe (including a special feature on making your own), and stone. A separate chapter on paving with these materials begins on page 56.

Once you've had a look at the how-to instructions in this and the following two chapters, turn to the projects section beginning on page 84. There you'll find ideas for everything from simple brick steps to poured-concrete retaining walls.

BASIC CONSIDERATIONS

Before beginning your wall, drop by your local building department and inquire about regulations that may apply to your project. These will specify how close to your property line you can build, how high you can build, what kind of foundation you'll need, whether or not the wall

Brick garden wall makes a natural backdrop for flowers. Flemish-bond wall is built with flashed brick and lightly raked mortar joints for a subtle interplay of color, light, and shadow.

will require steel reinforcing, and more.

Building to code

For some, the mere thought of trekking down to the building department in order to dive into a mass of red tape is intimidating. In reality, the building department can be a real asset to any do-it-yourselfer.

Building officials are usually pleased to help you with your project, and they can be of great assistance in steering you right. Their job is to enforce the building code, which is the set of regulations that specifies minimum standards for materials and workmanship. Constructing your wall — or any project — to these standards is cheap insurance; if you do, you can be assured of its structural integrity.

In the past, building departments were not especially concerned with non-load-bearing walls (walls that carry only their own weight), but in this era of litigation, things have changed. Many municipalities now require a building permit for masonry walls over 3 feet; some, especially in seismic areas, may also require that the wall be approved

by an engineer before you can obtain a permit. Be sure to check.

Walls lower than 3 feet and other small-scale projects will probably not require a permit, but it's a good idea to take a sketch to the building department anyway.

The how-to instructions that follow refer to walls 3 feet high or lower, and the discussions of steel reinforcing are intended only as introductions. If your project is more than 3 feet high, or if you need steel reinforcing, be sure to consult your building department.

Choosing your materials

Within the range of brick, block, and stone, there is certain to be a material that meets your needs. Here are some things to consider before choosing one.

Brick. With its wide range of hues and sizes, brick is usually the material of choice for garden projects where appearance is most important. Bricks are small, so working with them is not taxing; on the other hand, their small size means the wall will rise rather slowly. Bricks make

BRICK, BLOCK,

an attractive veneer over homelier materials, such as concrete block. This combination offers the advantage of brick's appearance with some of the strength, speed, and economy of block work.

Twelve easy steps to effective bricklaying begin on page 38. They are followed by instructions on laying corners, striking mortar joints, and more. To pave with brick, see pages 60–62.

Concrete block. This is your best choice for heavy-duty retaining walls and for projects where costs must be kept down. Where appearance is important, you can choose among many kinds of decorative surfaces—the slump block, which resembles adobe, is probably the most popular. Screen blocks allow you to build perforated walls that admit light and air.

Complete how-to instructions for building with both regular and screen blocks begin on page 46. Interlocking concrete paving blocks are discussed on page 57.

Adobe. Modern adobe blocks are handsome and much more durable than their ancient predecessors. Asphalt emulsions that waterproof and stiffen the blocks make the difference. You can use modern adobe just as you would brick, without fear of its disappearing in a rainstorm. Step-by-step instructions for working with adobe begin on page 50. You'll also find a special feature on making your own on page 51. To pave with adobe, see page 60.

Glass block. After a long slide into near-obscurity, glass block is staging a comeback. The blocks are laid up rather like bricks, but there are several special details of their installation which you should discuss with a professional. (See the Yellow Pages under "Glass Block, Structural.") Because of these complexities, we do not present how-to instructions in this book. For a photo of an outstanding modern application of glass block, turn to page 15.

Stone. If you have a good eye, a strong back, and lots of patience, you can use stone to build walls and other projects that have the handsome, rugged look only stone can give. Instructions for building stone walls begin on page 54. You'll find instructions for stone pavings on pages 60 and 63.

MORTAR

Mortar is the bonding agent, the "glue" that sticks masonry units together. Beyond this, it has several other functions: it seals out wind and water, compensates for variations in the size of masonry units, anchors metal ties and reinforcements, and provides various decorative effects, depending on how the joints are tooled, or "struck."

Mortar ingredients

Mortar recipes vary according to their intended use, but the ingredients are always the same: cement, lime, sand, and water. Consult your building supplier about the quantity of mortar you'll need for your project.

Cement. For general purposes, you should use Portland cement, a hydraulic cement (one that hardens in water). It is widely available.

Lime. Though lime weakens mortar somewhat, it is vital for making the mix workable. Hydrated lime is used in mortar; this material is caustic, and you should take care to avoid its contact with your skin.

Sand. Mortar sand should be clean, sharp-edged, and free of impurities such as salt, clay, dust, and organic matter. Never use beach sand; it is too rounded. Particle size should range evenly from about ⅛ inch to fine.

Water. If you can drink it, it's fine for mortar. Never use salt water or water high in acid or alkali content.

Here's how to measure

The most accurate way to apportion ingredients is to weigh them, but since this is rarely practical on the jobsite, masons usually go by volume. There are specific recipes for various types of mortar (see chart, page 35); in practice, looser, rule-of-thumb recipes tend to be used.

Once you're on your jobsite, ready to begin, you'll probably find it convenient to measure out your ingredients by the bag, bucket, or shovelful. The key is to be consistent in measuring so that your mortar will be the same from batch to batch.

Since mortar must be mixed in fairly small batches (large batches tend to harden before they're used up), masons often mix by the shovelful. A typical mix might be one shovelful of cement, one of lime, and six of sand. This is a good approxi-

ADOBE & STONE

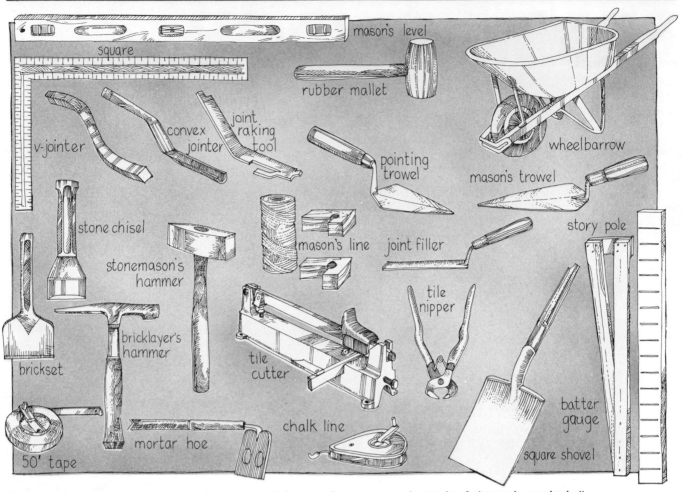

Tools for unit masonry include many you may already have and some you can improvise. A story pole, marked off for the thickness of a brick (or block) plus a mortar joint, can be made from scrap lumber, as can the batter gauge, used in stonework. An ordinary level used against a 2 by 4 can substitute for the extra-long mason's level. In fact, you may find that with a little ingenuity, all you need to buy are a good trowel and brickset.

mation of Type N, the commonest type for general use (see chart on facing page).

Mortar does become weaker as its lime and sand content go up, but it also becomes cheaper since the amount of cement (the most expensive ingredient) goes down in proportion. For most work around the home, Type N mortar has more than enough strength.

Most expensive of all is ready-mixed mortar, sold by the bag at building supply stores. Though brand recipes vary, most manufacturers produce a mortar similar to Type N. If your job is small, ready-mix is your best bet — the extra cost is usually offset by the convenience of not having to purchase, then measure, separate ingredients.

Mix it well

In mixing mortar, the dry ingredients are first measured out and mixed, either in a power mixer or by hand; then the water is added and mixed in. The amount of water needed cannot be specified in advance; it depends entirely upon the composition of the mortar and the absorption rate of the masonry units to be laid, factors which can vary according to the weather.

Ready for use, your mortar should have a smooth, uniform, granular consistency; it should spread well and stick to vertical surfaces, yet not smear the face of your work. Add water a little at a time until these requirements have been met. Experience will be the best teacher.

Power mixers. These are by far the best means to use if your job is large. Power mixers can be rented in a va-

riety of sizes to suit the work intended. When using one, add the sand, then the cement, then the lime, all with the mixer running.

Once the dry ingredients are well blended, start adding water. Be conservative; it's easy to add too much. The mixer should run for at least 3 or 4 minutes once all the water is added. Mix only enough to last you about 2 hours; more than that is likely to be wasted.

Mixing by hand. Small amounts of mortar can readily be mixed by hand. You'll need a wheelbarrow or mortar box and a hoe (see above).

As with the power mixer, mix the sand, cement, and lime well before adding water. Hoe the dry ingredients into a pile, make a hole in the top, and add some water; mix, then repeat the procedure. Repeat as often as necessary to achieve the proper consistency.

Another option is the "walk-along" mixer, which effectively streamlines the hand-mixing procedure. In essence, it is like a power mixer—the difference is that you are the motor. To use one, you load in the dry ingredients, walk the mixer along to blend them, then add water and repeat the mix-while-you-walk procedure. You'll need to be able to do this close to your worksite since the loaded mixer is quite heavy; be sure there are no intervening flights of stairs.

Mortar application

A master mason at work is a study in skill, speed, and concentration. If you can, try to observe one; you'll find the time well rewarded.

To lay bricks or other masonry units, you need to develop several skills with the trowel. You may not attain all the precision placement and joints of a professional, but after some practice you may find that the chief difference between you and the master mason lies in the speed of your work, not in its finished appearance.

Throwing a mortar line. For bricks and other masonry units of similar size, you'll need to learn to throw a mortar line—an even bed of mortar several bricks long. Here's how the pros do it.

Place one or two shovelfuls of mortar on a mortar board—a piece of plywood about 2 feet square will do. Load your trowel (an 8 or 10-inch trowel holds the right amount for brickwork) by slicing off a wedge of mortar and scooping it up. Give the trowel a shake to dislodge the excess.

Now comes the tricky part— throwing the line. Refer to the drawing to see how this is done. Essentially, it is a two-part motion: as you bring your arm back toward your body, you rotate the trowel, depositing the mortar in an even line about 1 inch thick, one brick wide, and 4 to 5 bricks long. It's a good idea to practice on the mortar board or other suitable surface until you get the knack.

Once the line is thrown, furrow it with the point of the trowel, using a

Mortar types

The following table lists ingredients by volume, using a single unit of cement as a starting point. As you read down the chart, you will see that the quantities of lime and sand are progressively increased. This makes the mix more workable and cheaper at the expense of some compressive strength. Type N is most commonly used.

Type	Cement	Hydrated lime	Sand	Chief characteristic	Best use
M	1	¼	3	High compressive strength	Below grade, as in foundations, walks, retaining walls—any place masonry will be in contact with water or damp earth.
S	1	¼-½	4½	High adhesion	Reinforced masonry, veneers—other applications needing extra bond strength and lateral resistance, such as in walls in windy areas.
N	1	½-1¼	6	High weather resistance	General-purpose use above grade, as in chimneys and exterior walls.

stippling motion. Take care to divide the mortar; don't scrape it toward you. The furrow ensures that the bricks are bedded evenly and will cause excess mortar to be squeezed out to either side as the bricks are laid.

Buttering. This is a self-descriptive term. The technique is used to apply mortar to the ends of bricks and other units. Mortar consistency is the key to successful buttering; mortar should be stiff enough not to drip, yet wet enough to stick.

Tempering. You can keep your mortar workable, or "well tempered," by sprinkling the unused mortar with a little water as necessary. There is a limit to this, however; never expect really stiff mortar to revive sufficiently for use. It's better to discard it and mix a smaller batch next time.

Striking mortar joints. Striking, or tooling, the joints compacts and shapes the mortar, contributing to the strength and weathertightness of the finished project.

Striking tools vary, depending on the type of joint desired; they may be as simple as a piece of wood or steel rod. After a section is completed, excess mortar is first cut off with the trowel, then the selected tool is drawn along the joint. Striking must be done before the mortar gets

too stiff; "thumbprint hard"—just hard enough to retain your thumb's impression—is about right. Popular mortar joints are illustrated on page 42.

Grouting. Grout is mortar that is thin enough to pour. It is used to fill cavities in masonry walls, such as the cells of concrete blocks (see page 45) or the space between the wythes of a brick wall (see page 37).

When grout sets up it locks a wall together into an essentially monolithic structure. Steel reinforcing is always secured in a wall by grouting.

To make grout, simply add enough extra water to your mortar so that it is liquid enough to pour.

Cleaning up. You'll soon learn that careful work is repaid at cleanup time. Keep mortar and dirt away from unit faces as you go; your finished project will look better, and you'll avoid a lot of unpleasant work with scrub brushes.

After joints are tooled and the mortar is dry enough not to be harmed by it, brush the surface with a stiff broom or brush to dislodge crumbs of dry mortar. Ideally, this will be all you need to do.

Should further cleaning be necessary, wash with a trisodium phosphate and laundry detergent solution. Try a half-cup of each in a gallon of water. Rinse well. If this doesn't work, more drastic means may be necessary (see page 80).

BRICK WALLS: HANDSOME & BASIC

Building with brick is pleasant work. The units are sized for easy one-hand lifting, and bricklaying takes on a certain rhythm once you get the hang of it. As with all building projects, planning and attention to detail are the keys.

The following section guides you through the design and construction of a freestanding brick garden wall. Retaining walls and brick veneer walls are special types. They are discussed on pages 88 and 86. Complete instructions for paving with brick can be found on pages 56–62.

Brick, pro & con

Bricks are used structurally in walls and pavings, both indoors and out. Their natural surface acquires a lovely patina with time, becoming more and more attractive with age.

A well-built brick wall is extremely strong in compression; that is, it resists crushing forces very well. It does not have much strength, though, in tension (stretching) and must be reinforced with steel where high tensile loads are expected: in structures such as a high wall in a windy area, or a retaining wall.

Brick has become relatively expensive, especially when compared with wood. Its expense, though, can be justified when you consider its beauty, permanence, and maintenance-free character.

DESIGNING FOR BRICK CONSTRUCTION

Tradition is your ally here. Bricklaying is a trade thousands of years old; it would be next to impossible to create a design totally without precedent. Once you begin to consider a masonry project, you'll find a world of pattern choices, some sure to be just what you're looking for. Here are some things to consider before beginning.

Running bond

Common bond

Flemish bond

English bond

Stack bond

Rowlock bond

Bond patterns. Over the years, masons have developed patterns, or bonds, in which brick can be laid (see drawing above). Be sure to check with your building department before making a final decision on bond pattern; if your wall will require steel reinforcing, some bonds may be more adaptable than others. For details on bond patterns, see pages 40–41.

Brick types. Because of the expense of shipping such a heavy material, brick is usually produced locally. No one type will be available all over the country, but there are two broad categories: **common brick** is that most often used for general building; **face brick** is used where strict uniformity of appearance is required. Within each category, you'll find a wide variety of colors and textures.

Brick sizes. Many brick are made in modular sizes – that is, the three dimensions are simple divisions or multiples of each other. The standard modular brick measures 8 inches long by 4 inches wide by 2⅔ inches high, so that two headers or three rowlocks will equal a stretcher (see Glossary on opposite page). This simplifies planning, ordering, and fitting. Those dimensions are nominal; they include the width of a standard ½-inch mortar joint, and actual dimensions of the standard brick are reduced accordingly. The

standard nonmodular brick measures 3¾ by 2¼ by 8 inches. It is also common for brick to vary somewhat from specified dimensions. To calculate the quantities of brick you'll need for your project, consult your building supplier.

Other sizes of modular brick may also be available. Jumbo, Roman, Norman, SCR, and Utility are just a few of those carried by large brickyards (see drawing). Cores (holes) are cast in some of those to reduce weight and improve mortar bonding—you might save money, since there is less clay in the brick.

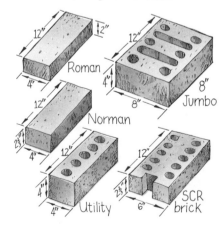

Footings. Poured concrete is the best material for footings. Codes usually specify a footing as deep as the width of the wall and twice as thick, with the wall centered above (see drawing below). Complete instructions for building a concrete footing can be found on page 69. Always be sure to check your local codes before proceeding.

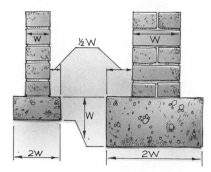

Reinforcing brick masonry

Walls over 2 feet high will need reinforcement. This may be as simple as a cap of header brick spanning the two wythes of a low wall (see Glossary at right), or as exacting as engineered steel reinforcing in a large retaining wall.

For higher freestanding walls, you have two ways to go: masonry or steel reinforcing.

Masonry reinforcing. Long walls, especially 4-inch, single-wythe walls, should be reinforced every 12 feet or so with pilasters (see drawings below). The pilasters are locked into the wall by overlapping the bricks in alternate courses, as shown.

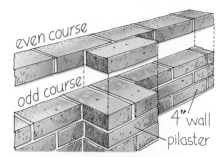

even course
odd course
4" wall pilaster

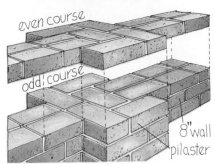

even course
odd course
8" wall pilaster

Double-wythe (8-inch) walls can also be built using headers. The header bricks running through the wall will help tie the wythes together. Common, English, Flemish, and rowlock bond patterns all use headers, but the alternation of header and stretcher courses differs from bond pattern to bond pattern. Double-wythe walls built in running bond can be partially reinforced by grout (thin, soupy mortar—see page 35) poured into the cavity between the wythes, but it's best to use steel ties.

Steel reinforcing. A simple way to add steel reinforcing to a grouted wall is to insert steel rods, called "rebar," into the grout once it has stiffened slightly. Insert the bars right down to the footing. Alternatively, place the bars in the footing when it is poured; then build the wall around them and grout it.

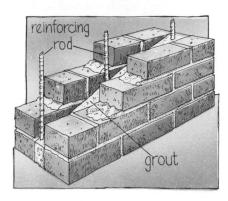

reinforcing rod
grout

Special steel ties of various patterns are made for reinforcing brick masonry. Two of the most common are shown in the drawing below. It is best to consult your building department to find out whether steel

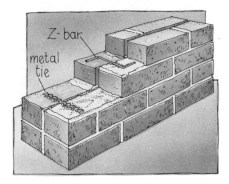

Z-bar
metal tie

A GLOSSARY OF MASON'S TERMS

Like all trades, masonry has its own vocabulary. Here are some of the terms you'll need to know.

Bed joint: Horizontal joint in a wall.

Bond: The method by which units are interlocked, the pattern made by units on the face of a wall, and the adhesion of mortar to units.

Closure: A whole or partial unit used to complete a course of masonry.

Course: Horizontal run of units in a wall.

Grout: Thin cement-sand mixture used to fill cavities in walls.

Header: Unit running across the thickness of a wall.

Head joint: Vertical joint in a wall.

Lead: Partial courses at the end of a wall that establish alignment.

Pointing: Repairing mortar joints by removing old mortar and adding new.

Raking: Removing fresh mortar from a joint; enhances the play of light.

Rebar: Steel reinforcing bar or rod.

Rowlock: Header unit turned on edge.

Stretcher: Unit running along length of wall.

Soldier: Brick or block standing with narrow face showing.

Sailor: Brick or block standing with broad face showing.

Striking: Cutting away mortar with the trowel; also, tooling mortar joints.

Wall tie: Metal reinforcing strip used to tie wythes together or attach veneers.

Wythe: Tier of masonry units one unit wide.

reinforcing will be required in your project, and if so, which kind.

Exact specifications and techniques for steel reinforcing are detailed in building codes; for further information, consult your building department.

HOW TO BUILD A BASIC WALL

Brick walls come in many shapes and sizes, as a glance at the color pages will show, but the freestanding garden wall is probably the most common and will give you the best introduction to bricklaying.

On these two pages you'll find out how to build a low garden wall. Corner layouts appear on pages 41 and 42; other freestanding walls are shown on page 43. Retaining walls are discussed on page 88.

Since walls more than a foot or so high must usually be two bricks thick, the wall shown on these pages is built in two tiers (wythes). It is in effect two 4-inch walls laid side by side and connected by header courses every sixth course. This bond pattern is called "common," or "American" bond. It's an all-masonry bond that is strong and easy to lay. Steel reinforcing usually is not required, at least up to the 3-foot limit permitted by most codes. Above 3 feet, you may need steel reinforcing, along with a building permit and an engineer's signature on the design. The instructions on these pages are only for walls up to 3 feet high.

If you wish to try a different bond pattern, turn to page 41. Since the common bond shown here uses both header and stretcher courses, you'll find the instructions for laying it adaptable to these other patterns.

Selecting your site. Careful location of the site will contribute to your wall's longevity. Choose a location where drainage is good and the soil is firm.

Avoid locating the wall near the root systems of large trees; the growing roots can exert a nearly irresistible pressure and may crack your foundation.

Finally, be sure to consult local codes for possible legal restrictions on your project's location.

Preparing to work. You can begin laying brick after the foundation has cured about two days. (For instructions on pouring a concrete foundation, see page 69.) Before beginning

work, distribute your bricks along the jobsite in several stacks. This will save you time later and will help you develop a rhythm to your bricklaying. Unless they are already damp, hose down the bricks several hours before beginning work; this prevents them from absorbing too much moisture from the mortar. Save any broken bricks for cutting. Have a hose or bucket of water handy for rinsing your trowel and other tools occasionally as you work. You'll need the water also to keep your mortar well tempered (see "Mortar," page 33). Mortar joints need to be tooled periodically as you work. See page 42 for instructions.

1 Marking the footing. Locate the outer edge of the wall by measuring in from the edge of the footing at each end so that wall is centered. Stretch a chalk line between the two points and "snap" it to mark your guideline.

2 Laying a dry course. Lay a single course of stretcher bricks (see Glossary, page 37) out along your chalk line the full length of the wall in order to mark bricks' spacing on footing. Allow ½-inch spaces for the head joints, marking them on the foundation with a pencil as you go. If necessary, adjust the head joint width to allow you to lay the course without cutting any bricks.

3 Laying the first bricks. Take up bricks from dry course, leaving markings. Throw a mortar line (see page 35) three bricks long and lay the first three bricks. Butter head-joint ends of the second and third bricks and place them with a shoving motion so that the mortar is squeezed out of all sides of the joint. Use rule, tape measure, or story pole to check the course for correct height, then carefully check that bricks are plumb and level, using trowel handle to tap them into place. Never pull on a brick, as this breaks the bond. Check head-joint thickness against your pencil marks and trim off excess mortar.

4 Beginning the backup course. Lay three backup bricks just as you did the first three. Use your level to check that the courses are at the same height in each wythe, and use a header brick to check overall width of the wall. Do not mortar the two wythes together.

5 Beginning the header course. Cut two ¾ bricks to begin the header course, then lay three header bricks.

6 Finishing the lead. Continue laying stretchers until the lead is five courses high, as shown. Note that the fourth course begins with a single header. Use your level as a straightedge to check that the lead is true on each of its surfaces. Now go to the other end of the foundation and build another lead, following steps 1–6.

7 Filling in between leads. Stretch a mason's line (see "A Mason's Toolkit," page 34) between the completed leads as shown, then begin laying the outer course. Keep the line about 1/16 inch away from the bricks and flush with their top edges; double-check bricks with rule or story pole, since the line will sag if the wall is long. Lay bricks from both ends toward the middle.

8 Building to the top of the leads. Butter the last, or closure, brick on both ends and insert it straight down. Mortar should be squeezed from the joints. Then move the mason's line to the back of the wall and lay the backup courses. Always use the line, level, and rule or story pole (see page 34) to check the accuracy of your work.

9 Going higher. To continue upward, build new five-course leads at each end of the wall, repeating steps 3–8. Keep a constant check on your work as you go, with rule, level, and story pole. Sight down the wall periodically to make sure it is true.

10 Planning the cap. You can cap your wall in many ways; the simplest is a row of header bricks on edge—rowlocks. Lay them out dry as shown, allowing for mortar joints. If the last brick overlaps the end of the wall, mark it at the point of overlap. Score and cut this brick on the line you have marked.

11 Laying the cap bricks. Throw mortar lines as shown and begin laying the cap. Each succeeding brick should be well buttered on its face, and you should keep a careful check on joint thickness as you go.

12 Finishing the cap. If you have cut a brick, "bury" it four or five bricks from the end, where it will be less noticeable. When you have laid the last brick, go over your work carefully with a level, checking the cap for alignment in all directions.

HOW TO CUT A BRICK

If you want really precise brickwork or are using extra-hard bricks, such as "clinkers," you'll need to rent a masonry saw or purchase a special blade for your circular saw. Softer bricks are easy to cut without a saw; just follow this procedure.

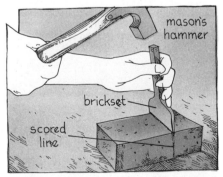

Score the brick, placed on sand or earth, by tapping with hammer on brickset. Go all the way around it. Cut the brick with a sharp blow to the brickset.

HOW TO BUILD CORNERS

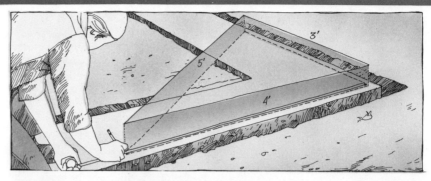

A wall with one or more corners is nearly as easy to build as a simple freestanding wall. Refer to pages 38–39 for basic bricklaying instructions. The drawings at right show a corner in common bond. You can adapt the instructions below to any of the other bond patterns (see facing page).

1 **Checking for accuracy.** After snapping your chalk lines, be sure they are absolutely square by using the 3-4-5 rule: Measure 3 feet along one line and 4 feet along the other. Now measure the distance between these two points. It should be 5 feet; if it isn't, adjust your lines. Using simple multiples of 3, 4, and 5—such as 6-8-10—will assure even greater accuracy.

2 **Starting the corner lead.** After making a dry run for the entire wall (see step 2, page 38), lay the first brick exactly at the corner, lining it up carefully with your chalk lines.

3 **Tailing out the lead.** Lay the remaining four lead bricks, checking carefully for accuracy. Masons call this "tailing out" the lead. Use the level to level and plumb the bricks and to check alignment, as shown. A steel carpenter's square will also help.

4 **Laying the backup course.** Throw mortar lines and lay the backup course as shown. Be careful not to disturb the lead course, and remember there is no mortar joint between these courses. Be sure the backup course is level with the first one.

5 **Starting the header course.** Take two bricks and cut them into ¾ and ¼ pieces, called "closures." Lay them as shown and complete the lead header course. This method can be used for other bond patterns with header courses, such as English bond.

6 **Completing the lead.** Now lay the leads for the next three stretcher courses. Note that each of these courses is the same as the first course, except the fourth course, which begins with a header. Check your completed lead for accuracy, then repeat these steps for any other corners in the wall. Now you're ready to finish the wall between leads, as shown on pages 38–39.

7 **Topping it off.** Plan and lay the row-lock cap as shown on page 23. Note that the cap course starts flush with one edge of the corner; lay these bricks first, then start the other leg of the corner by butting the bricks against the first ones. Use closure bricks as needed; see page 39 for instructions on planning the cap.

Running bond

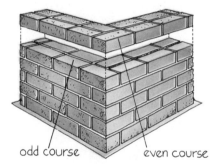

odd course   even course

English bond

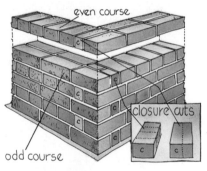

even course

closure cuts

odd course

Flemish bond

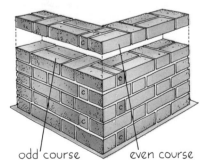

odd course   even course

Stack bond

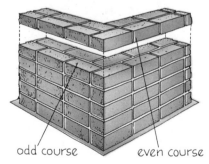

odd course   even course

OTHER BOND PATTERNS & THEIR CORNERS

Over the years many brick bonding patterns have evolved. The ones most frequently used are shown here. By referring to the instructions for laying the common bond (pages 38–41), you'll be able to use these patterns to achieve variety and distinction in your work.

Running bond is easy to lay, and is most often used for veneers and single-thickness partitions. Double thicknesses must be linked with metal ties (see page 37).

English bond, commonly used in England for structural work, forms very strong all-masonry walls. Header courses require closure bricks at corners, as shown.

Flemish bond alternates headers and stretchers in each course. It is both decorative and structural.

Stack bond is usually used for decorative effect in veneers. Because there is no overlapping, it is a weak bond and must be liberally reinforced if it is to be used structurally.

HOW TO FINISH MORTAR JOINTS

Finishing, or "striking," mortar joints can be left until the end, if your job is small. For larger jobs, though, you'll need to do this periodically as you work. Properly tooled joints are essential to ensure strong, watertight walls.

Mortar joints should be tooled when they are neither so soft that they smear the wall nor so hard that the metal tool leaves black marks. "Thumbprint hard" is about right, though experience will teach you best.

1 Striking the bed joints. Run the striking tool along the bed joints first. See below for a variety of joint treatments.

2 Striking the head joints. Now press or draw the tool along each head joint.

3 Cutting off the tags. Slide your trowel along the wall to remove mortar forced out of the joints (tags). When finished, restrike the bed joints.

4 Brushing the wall. Once the mortar is well set, brush the wall with a stiff broom or brush; this will eliminate much later cleaning.

TYPES OF MORTAR JOINTS

You can choose among several commonly used joint toolings. Here are the advantages and disadvantages of each.

1 Extruded joint results when mortar is allowed to squeeze out (extrude) from the joint as the brick is laid. It has a rustic appearance but is not waterproof; thus it is not suitable in rainy climates or where there is much freezing weather.

2 Flush joint is made in the course of ordinary bricklaying; excess mortar is simply cut away with the trowel. This is not a strong joint, since the mortar is not compacted, and it may not be waterproof.

3 Struck joint provides for dramatic shadow lines but tends to collect water. The joint is struck with the trowel, yielding some compacting.

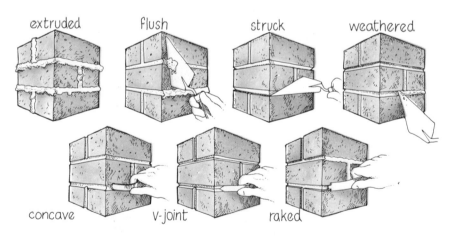

4 Weathered joint is struck from below with the trowel. It is the most watertight joint of all and is somewhat compacted.

5 Concave joint is made with a special jointer, as shown, or with a similar convex object. This joint readily sheds water and is well compacted.

6 V-joint is essentially similar to the concave joint in both strength and weather resistance. Use a special tool (shown), metal bar, or piece of wood to make this joint.

7 Raked joint casts dramatic shadows. Weather resistance is poor, though, and the joint is weaker than the concave and V joints. A joint raker is used for best results.

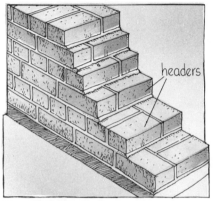
metal ties

Running bond

headers

Common bond

headers

English bond

full header
½ header

Flemish bond

MORE BRICK WALLS

In thick walls or thin, the basic brick will serve you well. Shown here are 12-inch-thick walls in four bond patterns, and two kinds of 4-inch-thick walls–the serpentine wall and the panel wall. One is older than the Republic and one features modern engineering; both overcome the usual height limits on thin walls, and both are distinctive.

All these walls can be built using the techniques presented on pages 38–42, but be sure to consult your building department before you begin.

Twelve-inch walls are commonly used where extra strength or thickness is needed, as in a high garden wall or a low, wide seating wall. As the drawings show, each of the most popular bonds can be laid 12 inches thick. A third wythe is added, with attention being paid to good principles of overlap bonding. The characteristic pattern is revealed on both sides of the wall, just as if the wall were only 8 inches thick.

cross section
brick foundation

Serpentine walls were used by Thomas Jefferson in his architectural designs 200 years ago, and they are still popular today. The distinctive, sinuous curve of the wall is actually an engineering feature; it helps the wall resist toppling and allows you to build thin walls higher than you could otherwise. The wall shown is only 4 inches thick. A brick footing is the easiest to use, but make sure this meets your local code.

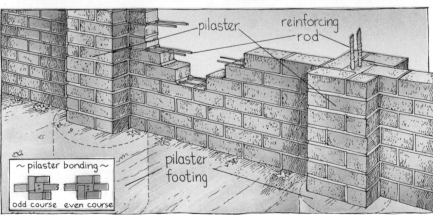
pilaster
reinforcing rod
~ pilaster bonding ~
odd course   even course
pilaster footing

Panel walls are a relatively new development; they must be carefully engineered. Steel is used to tie together a structure of brick panels and reinforced pilasters. Only the pilasters have foundations. Support for the panels (which are based right on the ground) comes from horizontal steel reinforcing that ties them into the pilasters. Panel walls can save much time and material. Engineering is critical, though; tables showing the reinforcing needed for various spans and configurations are available from the Brick Institute of America, 1750 Old Meadow Road, McLean, VA 22102.

CONCRETE BLOCK WALLS: STURDY & ECONOMICAL

For fast, inexpensive masonry wall construction, it's hard to beat concrete blocks. These rugged units make strong decorative and structural walls, and working with them is usually well within the capacity of the do-it-yourselfer.

Concrete block, pro & con

The large size of concrete blocks – 8 by 8 by 16 inches is standard – makes for rapid progress in building. Most freestanding walls can be built with only one thickness of block. This is in marked contrast to brick, and it's a real timesaver that helps compensate for the blocks' rather cumbersome size and weight.

Where extra strength is needed, the hollow cores can be easily filled with steel reinforcing rods and grout.

When a wall is heavily reinforced and grouted solid, the blocks become, in effect, permanent forms for what is really a poured-concrete wall. The result is a monolithic wall highly resistant to both tension and impact, two areas where the blocks alone are comparatively weak because of their small bonding surfaces and thin shells.

The blocks are available in an array of sizes and shapes; all except slump block have precise dimensions, making it easy to plan and design your wall. Since the only factor limiting accuracy is your ability to maintain a consistent ⅜-inch mortar joint, working with concrete block can be quite exact – more so than with brick, where sizes may be slightly irregular.

On the other hand, you need to be precise with blocks. Cutting them is hard to do accurately without a power masonry saw, so it's best to base the overall dimensions of your project on the block you will use.

Concrete block is usually much less expensive than brick or stone; it's also generally considered less attractive than these materials. This is

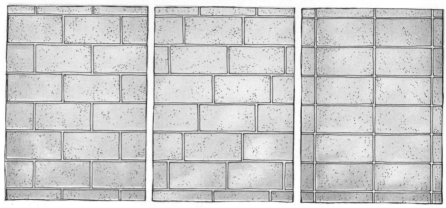

Basic bonding patterns for concrete block include running bond (left), offset running bond (middle), and stack bond (right). Stack bond requires extreme care in vertical joint alignment, as unevenness is easy to see.

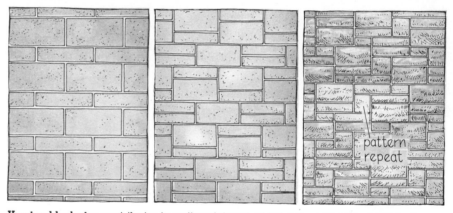

Varying block size contributes to pattern interest: at left, full and half-height blocks; middle, the same blocks in a pattern reminiscent of ashlar stonework (see page 53); right, four sizes of split-face blocks (see "Decorative block," opposite page) in another ashlar pattern.

no problem if you use the block wall as a core over which you apply brick or stone as a veneer (see page 87). Appearance can also be improved by the use of decorative block, or a combination of standard blocks, to give variety to the texture and bonding pattern of the finished wall (see drawings above).

DESIGNING A CONCRETE BLOCK WALL

Since concrete blocks are cast to such accurate sizes, you can plan with confidence. You'll need to pick the type of block first, then base the overall dimensions of the wall and its footing on the dimensions of the block. For more on footings, see page 67.

Remember that block dimensions are nominal; each measurement includes a standard ⅜-inch mortar joint. Also, check with your building department if the wall is to be more than 3 feet high. Codes may restrict your design above this height, and the instructions on these pages do not cover higher walls.

Steel reinforcing. Taller block walls and most block retaining walls will require steel reinforcing. This is easily done by placing the vertical rods in the foundation so that they extend up through the block cores, which are then grouted.

Special bond-beam blocks provide for horizontal reinforcing. Their cutaway webs allow the placement of reinforcing rods and grout. Since code requirements vary, and steel reinforcing is complex, be sure to check with your building department before attempting to design and carry out a reinforcing system.

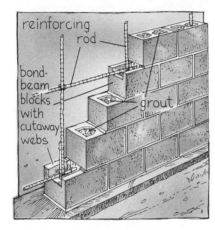

Kinds of concrete block

Like other masonry materials, concrete blocks are usually made locally. Thus, not all kinds shown on these pages and in the color section may be available in your area. Even the standard 8 by 8 by 16-inch block varies from coast to coast; in the West it has two cores, in the East, three. East, west, or in between, when you visit your local supplier you'll probably find two broad categories of block: basic and decorative.

Basic block. In addition to the 8-inch-wide standard size, blocks come in 4, 6, and 12-inch widths. You might consider these if your wall is to be very low or if you want a massive effect.

Two weights of block are available. Standard blocks are molded with regular heavy aggregate and weigh about 45 pounds each. "Cinder" blocks, or lightweight blocks, are made with special lightweight aggregates and weigh considerably less. Either type is suitable for most residential projects, so you can

make your selection on the basis of cost or your aching back, or both.

Whatever the size and weight of the standard block, a whole series of fractional units is likely to be available to go with it. In addition, the standard block itself will probably be found in at least stretcher and corner form (see drawing below). It's easy to see that with a little planning and care in assembly, you'll never have to cut a block.

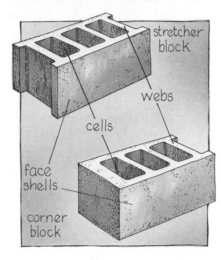

Decorative block. Most manufacturers make a variety of decorative blocks designed to produce surface patterns that catch the play of light, enhancing a wall's appearance (see drawing at right).

Sculptured-face units have patterns cast in relief on their face shells. These blocks can be combined in various ways to produce overall patterns in a wall.

When slump blocks are made, they go through a press that gives them an irregular, "slumped" appearance similar to trimmed stone or handmade adobe. Dimensions are somewhat irregular, in contrast to most other blocks.

Split-face units are actually broken apart in manufacture and resemble cut stone. Combining several sizes enhances their effect.

Screen or grille blocks are designed to be laid on end, usually in stack bond (see page 47). They form patterned screen walls that admit light and air while still affording some privacy.

With a palette this broad, the design possibilities are endless. Be sure to see what's made in your area before you start planning, though; few block designs are standard.

Estimating your needs

It's a good idea to make an accurate drawing of your proposed project, showing the actual number of blocks per course and the number of courses. Then it will be easy to see how many blocks you'll need.

For example, a wall 8 feet long would require six standard blocks per course; two corner blocks and four stretchers. For running bond—the best bond to use for concrete blocks—every other course would begin and end with a half block and contain five standard stretchers.

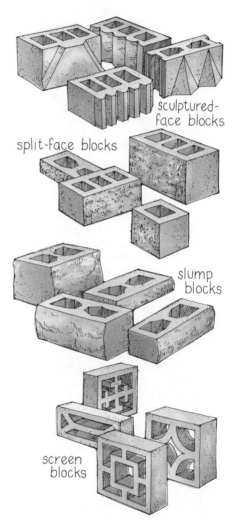

Since each course is 8 inches high, it's easy to figure the number of courses needed to attain a given height, and thus the total number of each type of block you'll need. It's a good idea to order a few extra for waste.

BUILDING A BLOCK WALL

Laying concrete block is much like laying brick. Running bond is most common, though stack bond, well reinforced, is also used frequently. Both patterns allow cores to line up, making steel reinforcing and grouting easy. Many other patterns are possible, especially if you combine several sizes (see drawings on page 44).

Laying concrete blocks

Before digging the footing trench (see page 38), lay out a dry course of blocks. Space them ⅜ inch apart, and try to plan the footing so that no block cutting will be necessary. (Should cutting be unavoidable, you can use the method described for bricks on page 39, or you can rent a special saw for the purpose.)

Now install the footing; complete instructions appear on page 69. After the footing has cured for at least 2 days, you can start assembling the wall. Mark the foundation as you would for brick work (see page 38). Note that the face shells and webs are thicker on one side of the block than the other. Always lay blocks with the thick side up; it gives more surface for the mortar bed. Proceed with the wall, using the instructions at right.

Mortar. Use the same mortar you would use for brick (see page 34 for mixing instructions). Keep the mortar a little on the stiff side; otherwise the heavy blocks may squeeze it out of the joints.

Do not wet the blocks prior to laying them, as you would bricks. The stiffer mortar and the lower rate of absorption of the blocks will keep them from absorbing too much water from the mortar.

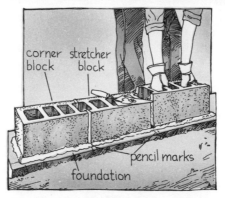

1 Starting the lead. Lay a 2-inch-thick mortar bed long enough for three or four blocks. Lay the corner block carefully and press it down to an accurate ⅜-inch joint with the foundation. Butter the ends of the next blocks, then place each ⅜ inch from the previous block. Use a level to align, level, and plumb the lead.

2 Completing the lead. Continue as for bricklaying (page 38), beginning even-numbered courses with half blocks. For maximum strength, you should mortar both face shells and webs, making full bed joints. When one lead is finished, go to the other end of the wall and build the second lead.

3 Filling in between leads. Lay blocks between the leads, keeping a careful check on ⅜-inch joint spacing. To fit the closure block, spread mortar on all edges of the opening and the ends of the block, then set it in place. Be sure to check alignment, level, and plumb frequently.

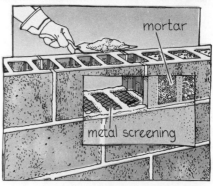

4 Capping the wall. You can make a simple cap by filling the cores of the top course with mortar. Cover the cores of the next-to-last course with ¼-inch metal screening or building paper before laying the top course. Be sure building paper does not interfere with the bond between face shells.

5 Decorative cap. Solid cap blocks in various thicknesses are available; these help give the wall a more finished look. Simply mortar them in place on full bed joints.

6 Finishing the joints. For best strength, a compacted concave or V-shaped joint should be used. A long sled jointer like the one shown is best, but a smaller jointer, or even a dowel, will do. Tool the horizontal, then the vertical joints. Finish by knocking off tags with your trowel. (See page 42.)

Bond-beam blocks cap the wall shown above. Reinforced and grouted, they substantially increase its strength. You can make your own bond-beam blocks by knocking the webs out of standard units but be sure to place ¼-inch metal screen or building paper under them to retain the grout. Note also how corner units overlap for strength.

Pilasters are used in long walls for extra resistance to tipping. Wall above is built of 12-inch standard units; 6-inch units provide the offset needed for the pilasters. Note how successive courses overlap for maximum bonding strength.

BUILDING IN EXTRA STRENGTH

It is often necessary to add some form of reinforcing to a concrete block wall. This is especially true if the wall will be exposed to high winds or possibly the impact of a car—or even a bicycle.

Most concrete block walls should have a *bond beam* cast into them at the top (see drawing, far left). Once the grout sets up around the steel, the top of the wall becomes a monolithic beam that greatly strengthens the wall.

Longer concrete block walls should incorporate pilasters. As the drawing at near left shows, this is easy to do, using full and half-width blocks.

HOW TO LAY SCREEN BLOCKS

Screen walls are fragile compared with other masonry walls, so be sure your foundation is rigid and your workmanship is very careful. The reinforcing shown is adequate for a 3-foot-high wall but be sure to see your building department if you plan to go higher; extra reinforcing will be necessary.

1 Starting off. Lay out the entire first course dry, planning for ⅜-inch mortar joints and marking each block's position with a pencil along a chalk line on the footing (see bricklaying instructions, page 38). Lay a mortar bed. Then lay the first course, buttering each block before placing it and checking for plumb and level as you go. Lay blocks atop the first course at both ends of the wall, and stretch a mason's line between them. Fill in the second course, using the line as a guide.

2 Reinforcing the joints. Add welded-steel joint reinforcement to the bed joints of every other course. (Joint reinforcement is available to fit all standard block sizes; see your building supplier.) Reinforcement should stop just short of the ends of the wall. Press the steel into the mortar bed before placing the lead block. As you fill in between the leads, throw mortar on top of the steel and lift the steel into the mortar before placing the block.

Concrete Block Walls

ADOBE: A TOUCH OF THE OLD SOUTHWEST

Nothing adds regional flavor to a garden like adobe. This simple mud brick was probably man's first manufactured building material, and it is still used today. Asphalt stabilizer now makes the blocks waterproof and has contributed to the gradual acceptance of adobe outside its native Southwest. When used in a garden setting, the rugged earthen blocks harmonize well with most kinds of plantings. But don't overlook indoor applications — sturdy hearths and fireplaces, for example, have been made of adobe for hundreds of years.

Adobe, pro & con

"Adobe" is a broad term. It can refer to adobe soil, an adobe block, even an adobe house. Thus it is possible to live in an adobe which is built of adobes which are made from adobe. Not only is the material economical, even its name is a model of economy.

Adobe is best used in open, generous gardens where the large size of the blocks will be in scale. A more confined space may be overwhelmed by them. The earth color of adobe looks good with natural woods and informal settings, but the blocks can be painted if you want a more formal feeling.

Modern adobe requires little maintenance beyond an occasional dusting. The stabilized blocks are waterproof, no longer subject to adobe's worst enemy — rain. However, if the soil is likely to shift or heave because of frost or other natural action, your wall may crack. A steel-reinforced concrete foundation and the use of steel in the wall itself will go a long way toward solving this problem.

Adobe is not especially good as an insulator; about 3 feet of it is necessary to provide insulation equivalent to a conventional, insulated, stud-frame wall. But it is very slow to transfer heat, and this quality makes it useful in solar and energy-efficient design. Since an adobe house heats and cools so slowly, it is possible to be cool while the sun is up and warm at night — when the stored heat of the day finally makes it through the wall to the interior.

Housebuilding is beyond the scope of this book, but the principle of slow heat transfer is a good one to remember when you're considering adobe for garden projects, fireplaces, and the like.

Adobe is readily available in the Southwest where the blocks are manufactured. You should check availability in your area, because added charges for shipping outside a plant's local area can more than double your costs. A possible solution is to make the blocks yourself (see box, page 51), though this will greatly increase your investment in labor.

Clay bricks are sized for easy handling, but this is rarely the case with adobe blocks, which may weigh up to 45 pounds each. Laying them can be quite a chore, especially when the wall reaches chest height. On the other hand, if you're in no hurry, you can pace the work so that your enjoyment in building the project is as great as your enjoyment of it once it's finished.

Adobe composition

The old-fashioned adobe block was made of a soil high in clay, with straw added to bind the block together and keep it from cracking. Modern adobe is quite different. Stabilized adobe is less than 15 percent clay; the asphalt stabilizer replaces much of the clay content and is the chief waterproofing agent. Straw is rarely used in modern adobe; here again, the asphalt takes its place.

Asphalt emulsions. These are the key to the success of modern adobe. Originally formulated for use in road work, any one of several available types may be suitable for stabilizing adobe mixtures.

An asphalt emulsion consists of very fine droplets of asphalt suspended in a liquid. Upon evaporation of the liquid in an adobe mix, the asphalt droplets are brought into close contact with the clay and soil particles. The asphalt coats each particle and fills the spaces between particles, waterproofing the block literally grain by grain. The stiffness of the asphalt also improves the compressive strength of the blocks.

Making your own adobe. It is a temptation to make your own adobe, since the process, however laborious, is simple. But to make a really strong adobe block takes some doing. Researchers at Fresno State College have come up with a reliable process for small-scale manufacture of adobe blocks, using road oil, soap, and soil. Refer to the box on page 51 for an outline of this method and an address for further information.

Sizes of adobe blocks

Adobe blocks come in several sizes. Blocks are generally 4 inches thick by 16 inches long, the width varying from 3½ to 12 inches. The most common block is 4 by 7½ by 16 inches, about the same as four or five clay bricks. Because the blocks are large, adobe wall construction proceeds quickly, and your efforts yield immediate results.

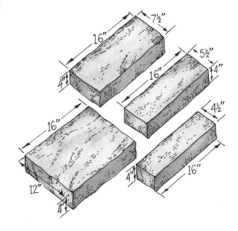

Laying adobe blocks

Laying up an adobe wall is similar to building with concrete blocks or bricks. Running bond is the pattern almost always used.

Because of their weight, you'll need a sturdy footing for adobe blocks (see page 69). Because the blocks are less resistant than bricks to flexing of the wall due to soil movement, the footing must be very stable. It should also extend 6 inches above ground level, to keep the blocks dry.

Tensile strength is added to the wall itself by steel reinforcing rods laid in the bed joints (see drawing at right). Be sure rods run continuously around any corners. Some walls will need vertical steel rods. The rods extend from the footing up through holes drilled in the blocks with a wood bit. Consult your building department for details on steel reinforcing.

Cutting the blocks to fit is easy. Use the procedure for cutting bricks shown on page 39. Or try an old hatchet—sometimes this works just as well.

Mortar. Adobe mortar is used in traditional adobe construction, but best practice today calls for Portland cement mortar, especially if you're using stabilized adobe blocks. Use a "leaner" (less cement) mixture than for bricklaying. The formula for adobe mortar is one part cement to two parts soil to three parts sand. Add 1½ gallons asphalt emulsion per sack of cement to waterproof the joints. The added soil gives the mortar an adobe color and "leans" the mix.

Ideally, the soil added to the mortar should be the same as that used to make the blocks. This is no problem if you live near the manufacturer—or *are* the manufacturer. Don't use clay soil—the type often called "adobe." Many clays are detrimental to mortar and to adobe blocks themselves.

Not only the proper soil, but also the asphalt emulsion is available from your adobe supplier.

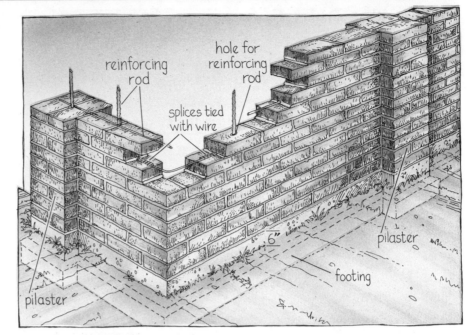

Anatomy of an adobe wall. Drawing above shows typical footing, reinforcing, pilaster, and corner details of an 8-inch-thick adobe wall. Footing extends 6 inches above the ground, is twice the width of the wall, and still wider where it supports pilasters. Half-inch steel reinforcing rods are set vertically in the footing and extend through holes bored in the blocks every other course. Steel rods laid horizontally in alternate courses are ⅜-inch thick; curved corner pieces are overlapped and spliced with wire (see page 50). Running bond is the pattern used for a simple and effective interlock at the corner.

How to order

If your local building materials yard doesn't carry a sufficient supply of adobe, they can order for you from an adobe manufacturer. Or you can order directly from an adobe manufacturer and have the adobes sent to you direct. Be sure to inquire about freight charges. You might also make your own arrangements with a hauling contractor.

How much to buy. Using the 4 by 7½ by 16-inch block to build a nominal 8-inch-thick wall, order as follows for each 100 square feet of wall:

Adobe blocks	200
Sand (cubic feet)	9
Portland cement (sacks)	3½
Asphalt stabilizer (gallons)	5½

The amount of reinforcing steel you'll need will depend on the height and length of the wall.

Finishing adobe

Adobe plaster covered with a coat of whitewash is the traditional finish for adobe. This gives a molded, monolithic look to the completed wall, and was necessary to protect old-fashioned unstabilized adobe from rain. But stabilized adobe requires no finish at all, and it's a good idea to postpone your decision on finish until you've seen how the completed project looks in its setting. Once you plaster or paint your wall, its rustic appearance is lost. Even if you later remove the finish, the molded texture of the original block will be gone.

But if painting is your choice, you'll find good results can be had with modern exterior latex (water-base) paints. Don't use oil-base paints; they may dissolve some of the asphalt stabilizer, allowing it to bleed through.

You can color the cement mortar by adding oxides as you mix it. See your dealer for available colors and proportions.

HOW TO BUILD AN ADOBE WALL

Follow these instructions for adobe walls up to 3 feet high. For specific reinforcing requirements, consult your building department.

1 Laying the blocks. Refer to "How to Build a Basic Wall," pages 38–39, for instructions on bricklaying. The method is essentially the same with adobe; only the scale is different. As shown at right, you use more mortar, and the blocks are much larger and heavier. Use a mason's line and work from leads, as you would for a brick or concrete block wall, maintaining a ½-inch mortar joint.

2 Filling in between leads. Drawing shows a block being added to the wall after a new lead has been completed. Reinforcing rods, shortened here for clarity, project from the new lead.

3 Reinforcing corners. Pieces of rod are prebent for corners, then spliced to the rod running along the bed joints. Standard practice calls for an overlap of 40 diameters; wire is wound around the rods to tie them together. For the ⅜-inch rod shown here, the 40-diameter overlap would be 15 inches. Consult your building department for specific reinforcing requirements.

4 Finishing the surface. Mortar joints can be tooled as for brick (see page 42), or struck off flush with the blocks for a characteristic Southwestern look, as shown here. Wall can be left natural or painted.

MAKING YOUR OWN ADOBE

It takes a lot of work, but if you have the space, time, and desire, you can make your own adobe blocks. Here are some things to consider:

• Don't take on a really large-scale project. Adobe blocks are very heavy. You may need hundreds for an ambitious project; at nearly 50 pounds each, that's a real tonnage.

• Allow enough time. It's time-consuming to make the blocks, and after they're finished, they must cure for at least three weeks in sunny weather.

• Test your blocks. If your soil mix is not just right, your blocks may crack. A laboratory must check your blocks (at your expense) before you can pass a building code or qualify for a loan on your project. More informal tests may suffice for small-scale projects.

All of these cautions mean that you really shouldn't try making your own blocks unless your project—a low garden wall or planter bed, for example—is not subject to code regulation.

The illustrations on this page are a general guide to making your own adobe blocks. They depict the steps used in a method developed at California State University, Fresno. For complete details on how to do it yourself, including standards for testing soils, emulsions, and finished blocks, write to: International Institute of Housing Technology, c/o Executive Vice President, California State University, Fresno, CA 93726.

Molds. Gang molds like those shown below can be real labor savers. It takes two to lift off the molds, but each mold makes several adobe blocks, speeding production. Any smooth lumber will do; the molds shown are made from 2 by 4s. The cleats shown, made from 2 by 2s, aid in lifting the molds.

Mixing the adobe. First the soil goes into the mixer, along with just enough water to make it slightly plastic. Then the emulsion is added, followed by enough water to bring the mix up to molding consistency. Mixing must ensure dispersion of the asphalt; the simple tumbling action of most cement mixers is not adequate, so a blade-type mixer is used. Hand mixing with a hoe also works well.

Molding the blocks. First, the molds are soaked in water. Just before molding, they are removed and placed on a hard surface covered with paper. A wheelbarrow and a couple of friends aid in the transfer of the mud to the molds. Once the

molds are full, the mix is tamped into all corners by hand, then screeded off.

Immediately after molding, the molds are lifted off and scrubbed clean with plenty of water. If this is done promptly, the next blocks will not adhere to the molds.

Drying and curing. On hot, windy days, a layer of paper placed over the blocks just after molding will prevent the cracking that can occur if the surface is allowed to dry too

quickly. After one day's drying, the blocks can be stood on edge for faster curing. Blocks are ready for use after about 3 weeks of curing.

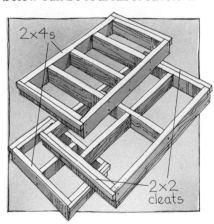

2×4s

2×2 cleats

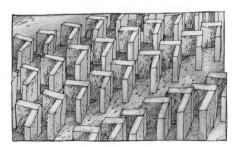

STONE WALLS: ELEMENTAL BEAUTY

Natural and enduring, stone has been used as a building material for thousands of years. Much of what we know of past civilizations comes from stone structures that have long outlasted the cultures that built them. Virtually no other material is as durable as stone—and stone, of all masonry materials, lends an unmatched aura of permanence to the finished structure.

Stonework ranges in appearance from the casual look of countryside rubble walls to the kingly formality of exactly fitted ashlar masonry. In the next four pages you'll find complete instructions for building 3-foot-high stone walls; stone pavings appear on pages 60 and 63. Before you choose to work with stone, though, you should consider some relevant facts.

Stone, pro & con

Today stone is an out-and-out luxury. Stone is "labor-intensive"—it takes a lot of time and effort to quarry, trim, haul, and store it. Even when found lying on the ground, it's a difficult material to handle and position. This didn't matter to the ancients; labor was cheap. Now, because of its expense, stone is reserved for applications where character and appearance are important.

For the garden, stone is never out of place. Both freestanding and retaining walls can be built either mortared or dry (unmortared). A nice feature of the latter is that you can place earth and plants in the joints as you build, blending the wall into the garden.

Stonecraft can be difficult work; it is probably the most laborious of the masonry techniques. Almost all forms of stone are denser, heavier, and larger than brick, and the irregular shapes of all but the most exactly trimmed ashlar stones (see

"Ashlar masonry," at right) make it difficult to keep large walls plumb and true while at the same time maintaining good bonding; they also pose a more demanding design challenge. Heavy stones are likely to crush their mortar beds—and your fingers, if you aren't careful.

Still, these problems can be avoided if you work slowly and carefully. The poet Robinson Jeffers, who was also a noted stonemason, built the large tower shown on page 19 almost without assistance. It took him more than 4 years, but the work gave him great satisfaction, and the result has a presence and majesty matched only in his poetry.

DESIGNING FOR STONE MASONRY

To begin, you'll need to choose a kind of stone and a method of laying—choices that determine the overall design to a large degree. Dry-laid walls will need more "batter" (inward slope) than walls built with mortar. So will rubble walls (as opposed to ashlar), whether dry-laid or mortared. Also, if you choose a mortared wall you will first have to construct a concrete foundation (see page 69); dry-laid walls may be built directly on the ground.

Kinds of stone

Stone can be divided into three broad categories according to the way it was formed: igneous, sedimentary, and metamorphic.

Igneous rock. Formed deep in the earth from molten magma, igneous rock is usually the hardest, most intractable kind you can buy. It includes granite and basalt, both durable, heavy stones.

Sedimentary rock. This rock was put down as layers of sediment, and it differs widely in texture, color, and composition from place to place. Sandstone and limestone are the

most common varieties. Sedimentary rock often splits easily along the sediment lines, making it useful for paving and ashlar stone.

Metamorphic rock. Marble and slate are the best-known metamorphic stones. As the name implies, they have been changed in form. Enormous heat and pressure in the earth act to create metamorphic rock from other kinds, sometimes completely obscuring the original stone. This is usually a durable material.

Types of stonework

There are two broad classes of stonework—rubble and ashlar. Between these extremes you'll find all sorts of "roughly squared" stonework, where some trimming of the stones has been done. Cobblestones are an example.

Rubble masonry. In this type of stone masonry, the stones used are often rounded from glacial or water action and include river rock and fieldstone. The rocks are often of igneous origin—hardheaded granite and basalt. Since they're tough to cut, it's usually easier to search for the right-sized rock.

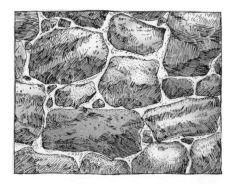

Rubble stonework is built up without courses. You fit the stones by paying attention to overlap bonding—stones overlapped so the vertical joints are staggered—and the pleasing arrangement of different sizes.

Because of the irregular spaces between the stones, much more mortar is required than for ashlar construction.

Rubble stone is usually the cheapest available—sometimes it's free. The famous New England stone walls were built by farmers in the course of clearing their fields; though secondary, the wallbuilding not only handled the disposal problem but also provided enclosure.

Ashlar masonry. Fully trimmed ashlar stone can be nearly as easy to lay as brick. The flat surfaces and limited range of sizes make coursing possible and require less mortar than for rubble work. You can get a less formal effect by avoiding regular courses and still take advantage of the labor-saving qualities of the cut stones.

The stone used is usually sedimentary in origin—sandstone is probably the most common. The stratification of this stone makes it easy to split and trim. It is softer and less durable than igneous rock, but this is of little concern in nonstructural applications. When an igneous stone, such as granite, is cut and trimmed for ashlar masonry, costs are likely to be quite high.

Bonding in stone walls

Here, as in brickwork, the principle of overlap is important. As you work, be sure that vertical joints are staggered; there should always be an overlap with the stones above and below.

Freestanding walls are usually laid up in two rough wythes with rubble fill. *Bond stones*, equivalent to headers in brickwork (see page

BONDING PATTERNS IN STONE WALLS

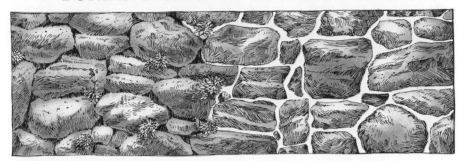

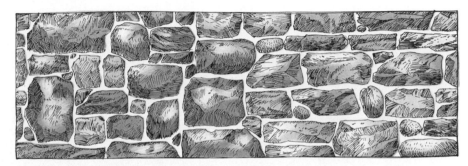

Untrimmed rubble can be laid "dry," as at left, or with mortar, as at right. Crevice plantings give the dry wall a natural look. Vertical (head) joints are always staggered, for maximum strength.

Roughly squared stones get some trimming at the quarry. Laid without regular courses (left), the effect is similar to a mortared rubble wall. Laying the stones in regular horizontal courses (right, gives a look resembling ashlar stonework.

Ashlar stone can be laid either without regular coursing (left) or with coursing (right). Head joints are always staggered. Avoiding regular courses gives a more rustic effect; including them makes the wall more formal-looking.

37), run across the wall, tying it together. You should use as many as possible—at least one for every 10 square feet of wall surface.

Batter. Most stone walls should slope inward on both surfaces. This tilting of the faces is called "batter" and helps secure the wall, since the faces lean on each other.

The amount of batter depends upon the size and purpose of the wall, the shape of stones used, and whether or not the wall is mortared. A good rule of thumb is 1 inch of batter for each 2 feet of rise—more if

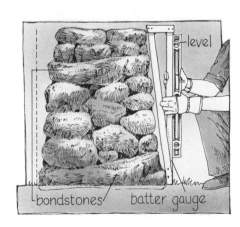

the stones are very round, less for well-trimmed ashlar. Mortared walls can make do with less batter, sometimes none. To check your work, make a batter gauge like the one in the drawing on page 53, and keep its outer edge plumb as you go.

HOW TO BUY STONE

If you figure the cubic volume of your wall, your dealer can figure the quantity of stone you'll need. Some dealers sell by the cubic yard, simplifying your order; others sell by weight. To find the volume of your wall, multiply its height, width, and length.

Rubble stone will have a greater volume per ton than trimmed stone; this is because of the voids between the rocks. Loaded in a truck, rubble stone might run as much as three parts rock to one part void—25 percent air. Remember this if you haul the stones yourself. Once you begin fitting the stones into a wall, you'll find their volume considerably reduced.

Try to inspect before you buy. Stones should harmonize in color and texture and should show a good range of sizes. For best effect, the face area of the largest stones should be about six times the face area of the smallest.

Trimmed stones are ordered by width and thickness, and you should specify the upper and lower limits of length.

HOW TO BUILD A DRY STONE WALL

A freestanding dry stone wall makes an attractive garden divider. The stones hold each other in place by weight and friction, so construction and maintenance are simple. If you like, earth and small plants can be placed in the joints as you proceed.

Unless your stones are very flat and square, plan on a battered wall (see page 53) no higher than it is thick at its base. Very round stones will require so much batter that the resulting dry wall may be no higher than a third of its thickness.

Even in severe frost areas, dry walls are built on very shallow foundations. Since the wall is flexible, frost heaves tend to dislodge only a few stones, which can be easily replaced.

Do save your largest stones for the foundation course; this will spare your back and make the wall stronger.

Cutting stone. To fit an awkwardly shaped stone, a certain amount of trimming can be done simply by hammering at the stone with a stonemason's hammer. If you need to cut a stone, a stone chisel or brickset will work (see page 34). Score a line completely around the stone, then drive the chisel against the line to break the stone apart. Try to work with the natural fissures in the stone, and always wear safety glasses.

1 First course. Lay the foundation stones in a shallow trench; this will help stabilize the wall. Begin with a bond stone, and then start the two face courses. Fill the center with rubble.

2 Second course. Lay stones on the first course, being sure that vertical joints do not line up. Stones of each face should tilt inward toward each other. Use your batter gauge and level on sides and ends to maintain proper slope. Again, fill the center with rubble.

3 Additional courses. Continue in the same manner, maintaining an inward tilt so that gravity will hold the wall together, and placing bond stones every 5-10 sq. ft. Use small stones to fill large gaps. If you tap them in with a hammer, the wall will be stronger. But don't overdo it—driving them in too far will actually weaken the structure.

4 Top course. Save your flattest, broadest stones for the top. If you live in an area with severe freezing, consider mortaring the cap; this will drain water away from the wall and help prevent frost damage.

HOW TO BUILD A MORTARED STONE WALL

A well-built mortared stone wall projects an image of solid strength and stability—an image that will be more than skin-deep if you observe the principles of good building outlined on this page.

Virtually any kind of stone can be mortared into a stable wall. Bear in mind that the quantity of mortar will increase sharply with progressively rounder stones. A wall 2 feet thick can safely be built without batter to a height of about 5 feet. Many municipalities do require an engineer's signature and a building permit for walls over 3 feet, though, so be sure to check. (The following instructions cover walls 3 feet high or less.)

The footing

A concrete footing is best. It should extend below the frost line; in frost-free areas, a trench about 12 inches deep is sufficient for a 3-foot wall. A gravel bed laid in the bottom will improve drainage, further stabilizing the wall. The idea is to absolutely ensure against movement, so that the inflexible wall will not crack. Leave at least 10 inches for the concrete, and plan to bring it to within an inch or so of ground level. The foundation should extend about 5 inches beyond the wall's edge on all sides, and it's a good idea to place a single ½-inch (#4) steel reinforcing rod about 3 inches up from the bottom. See page 69 for instructions on building a concrete footing. Let the footing cure for 2 days before beginning your wall.

Mortar for stonework

Because of its large joints and voids, a rubblestone wall may be as much as one-third mortar. To plan for this, lay up a small section of the wall, note the amount of mortar used, and use this as a guide for the rest. It's not a bad idea to use this method even if you're working with well-trimmed stone. Every stone wall is a special case; laying a sample section will be your best guide to the amount of mortar you'll need.

The mortar formula for stonework is richer than that used for brick: one part cement to three or four parts sand. You can add ½ part fireclay (available from your building supplier) for workability, but don't add lime (or mortar cement, which contains lime) because it might stain the stones. Keep the mortar somewhat stiffer than for brick.

1 **First course.** Your stones should be clean and dry (dirt and moisture interfere with the bond). Lay a 1-inch-thick mortar bed for the first bond stone and set it in place. Continue as for a dry wall (facing page), first mortaring the faces, then filling in the center with rubble and mortar. Pack the head joints with mortar after setting the stones.

2 **Additional courses.** For each subsequent course, build up a mortar bed and set the stones in place just as you did to begin. Work slowly, dry-fitting stones before throwing down the mortar. You can save mortar by filling large joints with small stones and chips. Check alignment and plumb or batter as you go (see drawing on page 53).

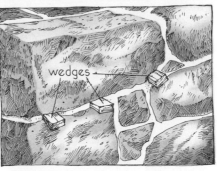

3 **Wedges for heavy stones.** Very large stones may squeeze out all the mortar in their bed joints. To preserve joint spacing, support them on wooden wedges. After the mortar is stiff, you can pull out the wedges and pack the holes with mortar.

4 **Raking the joints.** After you have laid a section, use a piece of wood to rake out the joints to a depth of ½ to ¾ inch. Deeply raked mortar joints enhance the play of light and shadow on the face of the wall. Ashlar stone walls can be tooled as for brick (see page 42).

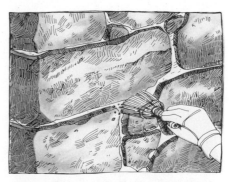

5 **Cleaning up.** Spilled mortar should be wiped from the face of the stone with a wet sponge as you work. After the mortar joints are tooled, use a broom or brush to remove crumbs of mortar. Once the mortar has dried, wash the wall with clear water. If this doesn't work, try soapy water and a clear rinse. Avoid using muriatic acid (see page 80) and steel brushes; these can mar the stone.

PAVING WITH

Every home needs some paving. You may want a patio suitable for lavish entertaining, or only a modest walkway alongside the house so you can drag out the trash cans on collection day. Whatever the need, some form of masonry, either poured concrete (see page 64) or unit-type, is likely to be the answer.

Unit masonry—brick, block, stone, and tile—usually offers the easiest approach. Preparations are usually simple; building can proceed on a weekend-by-weekend basis, and the results can be truly handsome.

Unit paving will probably be more expensive than poured concrete for very large areas, but for small to medium-size jobs, working with units can really pay off. The simplicity of laying brick in sand (see page 60), for example, may more than offset its cost when you consider that you are not contending with a concrete truck, can work at your own pace, and are rewarded with a beautiful, durable surface you can be proud of.

Today's paving units offer you a wide range of effects. As the drawings on the facing page show, your choice of material has a dramatic impact on the "feel" of your finished project. After you've established the location and scope of your project, consider your materials carefully; once they're down, they're likely to be down for a long time.

Some basic considerations

Each paving material has its strong and weak points; qualities that make a material good for walls may make it a poor choice for paving. For example, the rugged surface of natural stone that makes it so beautiful in walls can cause uncertain footing on a patio. Here are some things to think about:

Surface texture. Smooth, shiny surfaces can be slippery when wet, and rougher ones too absorbent for use near a barbecue. Smooth surfaces are best for dancing; games require surfaces with more traction. You'll need a hard-wearing surface if furniture will be dragged across it, but a softer one will do for foot traffic.

Appearance. Consider your taste in color, texture, pattern, and reflective quality. Do you want to match an interior floor or create a dramatic contrast? Dull surfaces mean less glare on the sunny side of the house; shiny surfaces can catch the light on the shady side.

Maintenance. How will you clean the paving? Most surfaces can be simply hosed down or swept, but some may show dirt more than others and need attention more often. Large mortar joints may trap debris, yet shed water well. Sanded joints are easy to maintain but may allow weeds to grow through.

Durability. Try to anticipate wear and tear from both climate and use. On stable soil, brick in sand is a permanent paving, but in areas with extreme freezing, you may get awfully tired of reworking the bricks every spring.

Cost. Consider more than just the material itself. Labor will vary, and hidden costs may lurk in such things as drainage, unstable soil, or special construction methods.

Application. Consider how to get your materials to the jobsite; if you have to hand-carry everything, you may not want to use really heavy blocks. Schedule your work for the best season, and try to anticipate the amount of help you'll need—also the amount of inconvenience you're willing to put up with as the work is proceeding. It's not a bad idea, by the way, to double your time estimate.

YOUR CHOICE OF MATERIALS

After giving thought to these basic considerations and narrowing your range of choice, you're ready to consider each material on its merits.

Brick: Warm & traditional

A brick path or patio is a gracious thing. As the drawing shows, this simple form lends itself to a broad range of paving patterns, and there are colors and textures of brick compatible with almost any home.

running bond   jack on jack

basket weave   herringbone

diagonal herring-   basket weave/
bone / jack on jack   grid system

The compact units make for easy work, and most kinds are easy to cut with simple tools (see page 39). The uniform size makes estimating and ordering easy, but because bricks are small, it will probably take 2,000 or more to pave an average patio.

Bricks are solid and durable, qualities that tend to offset their expense. Though a hard winter may cause a brick-in-sand surface to buckle, repair is easy: you simply pick up the offenders, relevel the sand bed, and re-lay the bricks.

The rough surface of common brick provides traction and reduces glare. The surface is porous, readily absorbing water. As the water evaporates, it cools the air and makes the surface cool underfoot. Unfortunately, it will just as readily

BRICK, BLOCK,

Brick laid in running bond gives patio a simple, traditional surface that harmonizes with the low seating wall.

Varying the pattern divides the single expanse and defines areas.

Mellow adobe, with its rounded, massive form, looks good with crevice planting.

Tile paving gives a more formal effect and a smoother, more reflective surface.

Rugged stone resists stains and scratches, gives a roughhewn effect.

Interlocking pavers are durable. Special edging pieces make cutting unnecessary.

absorb oil, grease, and paint—all of which may be hard to remove.

Interlocking concrete pavers: European immigrants

Concrete pavers, in their dozens of cleverly interlocking shapes, have long been used abroad for everything from patios to roadways. They are now made in this country and are finding increasing use in residential work.

Interlocking pavers are a logical industrial descendant of the old-fashioned, labor-intensive cobblestone. They are made of extremely dense concrete, pressure-formed in special machines. Laid in sand with sanded joints, they form a surface more rigid than bricks. No paver can tip out of alignment without tak-

ing several of its neighbors with it; thus the surface remains intact even under very heavy loads.

A special variant, the turf-retaining block, is designed to carry lighter traffic while retaining and protecting ground cover plants. This creates the possibility of grassy patios and driveways, and side yard access routes that will stand up to wear.

Pavers are laid in sand, just like brick (see page 60); alignment is nearly automatic. Cost varies from one make to another, but usually it's about the same as for brick.

Regular (non-interlocking) rectangular concrete pavers are also available. These are also used just like brick, but can be set in either sand or dry mortar.

Adobe: Regional character

It's hard to match adobe for friendly, rustic charm. Stabilized blocks con-

taining asphalt emulsion are waterproof and nearly as durable as brick.

Adobe paving blocks are available in square or rectangular sizes. They are usually laid in sand with one-inch sanded or dirt-filled joints (see page 60). The large joints help make up for size variations, improve drainage (always important with adobe), and allow for crevice planting, which adds to adobe's rustic appeal.

Adobe is cheaper than brick if you live near the source of supply. Otherwise, the cost of shipping may drastically raise its price.

Tile: Rustic to elegant

Tile gives an extremely smooth, durable outdoor surface. It can be as

STONE & TILE

hard as the toughest stone, and like stone, it resists abrasion and soiling.

Tile can be laid in sand or dry-mortared but is best mortared over concrete slabs and wood decks (see page 63). It's lighter than brick and no more difficult to work with if you don't have to cut it. Cutting requires special tools.

Tile usually costs more than brick, and the slab or decking base can drive the price up further. The finished surface is likely to be slippery when wet unless the tiles are quite rough.

Typically, tiles are a foot square or less in size. Most outdoor varieties are unglazed and take their color from the clay itself. This may range from gray to brick color. Glazed tile can be used outdoors but should probably be reserved for borders where people won't slip on it.

Stone: An old standby

Flagstone paving used to be more common than it is today; it still provides one of the toughest outdoor surfaces available. The subdued colors and irregular shapes greatly enhance most outdoor settings.

Flagstone is many times more expensive than brick but has some definite advantages. It is one of the few materials that can be laid directly on stable soil as well as in sand or mortar (see page 60). The large size and weight of the stones add to their stability, whatever the method of laying. Cutting and trimming are done with a brickset or stonemason's chisel and hammer—about the same as for brick (see page 39). Flagstone and slate are also available precut to rectangular shapes.

Some alternatives. River rock and fieldstone offer alternatives to the high cost of flagstone. These water-worn or glacier-ground stones produce a rustic, uneven paving that makes up in charm what it may lack in smoothness underfoot. You can also make your own "stones" of concrete (see page 76).

PREPARING THE BASE

The first step to stable masonry paving is a well-prepared base of firm, well-drained soil.

Drainage. Any time you pave an area you affect its drainage. Water will tend to run off even the most porous paving, such as brick in sand. Unless the area to be paved slopes naturally, you must grade it before paving so that runoff will not collect where it will cause problems—against a house foundation, for example. You should provide a pitch of at least one inch in 10 feet.

Try to avoid sending the runoff toward any place that is already boggy during a rain because it will only make the problem worse. Refer such a problem to a landscape contractor or architect.

Grading. Once you know which way to send the runoff, you are ready to start grading. Usually, this means you will be digging out the area to be paved. Plan to avoid filling and tamping; tamped earth is never as firm as undisturbed soil, and it will inevitably settle, taking your paving with it.

To grade the area, first stake it off in squares of 5 or 10 feet. Use a taut string to align stakes. Determine your preferred level for the finished paving surface and mark it on one stake. This should be at or above existing grade. Attach a chalk line at this point and stretch it along a row of stakes. Level the line (using a regular level or a line level, designed to hang on the chalk line) and snap it to mark the stakes. Repeat for all stakes.

Now adjust your marks to allow for a pitch of 1 inch in 10 feet (see Drainage, above). Plan to have the low side of the paving end at or above the natural ground level, not below it. You can do this by measuring down on the stakes at the low end or up on the stakes at the high end or both. Using a chalk line, mark the stakes again.

Excavate below your marks a distance equal to the paving thickness plus the thickness of the setting bed. As mentioned above, avoid filling, but if you must, wet and tamp the soil several times. Now you're ready to install edgings (see next page).

YOUR CHOICE OF METHOD

Masonry units can be laid in sand and by the dry or wet-mortar method. Each technique has

its advantages—and a few drawbacks.

Laying masonry units in sand

This is a simple method that yields surprisingly sturdy results. Typically, once the surface is graded, a strong edge is built, either with mortared masonry, poured concrete, or lumber (see facing page). A sand bed is then poured in and leveled, and the units are placed tightly together, filling the area. (Half-inch joints may be left between units, but this won't be as strong.)

The secret of the strength of this paving lies in the last step. Sand is swept into the narrow joints between units, where each grain acts like a tiny wedge to lock them together. An occasional resanding improves the wedging effect; traffic moves the units slightly, forcing the sand ever deeper into the joints and the whole pavement ever tighter against its restraining edges. If the edges hold and the ground doesn't move, this paving is simple and permanent. You can expect some settling over time, so it's a good idea to build a little high to allow for it. Occasional weeds in the joints can be kept down with a contact weed killer, or you can lay plastic sheeting between the bricks and the underlying sand.

Sand-bedding the units provides a flexible surface that allows for easy repair should tree roots or freezing weather cause it to buckle. Also, if a unit is damaged, it is easily replaced if laid in sand.

Mortared paving, dry or wet

Added stability is gained when you work with mortar (see page 34 for mixing instructions). Brick laid in wet mortar over a concrete slab (see page 62), which is in turn laid on a gravel base, provides the best protection against frost heaves and weed invasion. The dry mortar method provides some of this permanence with a lot less effort.

HOW TO BUILD EDGINGS

Strong edgings are the key to a secure pavement, especially if it's laid in sand (see page 60). An edging borders and retains the paving units, keeping them from moving. Shown here are five basic methods of building edgings—one of them should meet your needs. At right is a simple wooden edging. Below are masonry edgings, with and without concrete. Install edgings when you've completed grading (see opposite page). Complete the edgings before beginning to pave.

Brick-in-soil edgings are easiest to construct. The earth must be really firm and capable of holding the bricks securely.

In the drawing below, the earth has been cut away and a row of "sailors" (bricks standing side by side) has been installed. The full length of the bricks is buried to ensure against tipping.

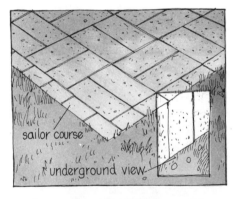

sailor course
underground view

The next drawing shows a variation. Here, the sailor course is titled 45°. This gives a notched effect at the edge and allows the pavement to

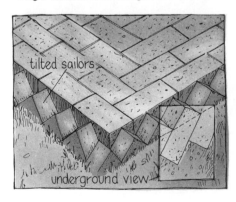

tilted sailors
underground view

area to be paved
allow room for cutoff
support joint with stake
cut off stakes

Three steps for a strong wooden edging: (left) after grading, stretch a guideline between two stakes to mark edging height; (middle) drive in stakes and nail on the edging (use 2 by 4 redwood, cedar, or pressure-treated lumber); (right) cut off stakes at an angle, then fill in to outer edge with soil.

rise above grade, at the same time keeping as much of the brick edging underground as possible.

To install either of these edgings, simply dig a narrow trench (if necessary after grading) and place the bricks. Level the tops with a bubble level as you go. Pack earth tightly against the outer perimeter of the bricks to secure them.

"Invisible" edgings rely on a small concrete footing to hold them in place. This is a strong type of edging that is effective with brick-in-sand paving and adaptable to interlocking concrete pavers, regular paving blocks, and other units.

The concrete is poured (see page 67) between temporary form boards set one brick length apart. A special screed (see drawing below) levels the concrete one brick-thickness below the top of the form. (The concrete should be about 4 inches thick.) As the screed is moved along, bricks are placed in the wet concrete. A few taps with a rubber mallet helps to set them.

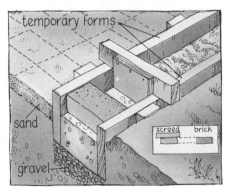

temporary forms
sand
gravel
screed   brick

The next day, the form can be removed. After the concrete has cured, the completed edging is used as a guide for a sand-leveling screed and the brick-in-sand paving is begun.

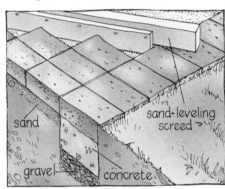

sand
gravel   concrete
sand-leveling screed

Concrete edgings are done in a manner similar to "invisible" edgings. Below, the concrete has been screeded or struck off flush (see page 69) with the top of a temporary form, creating a curb for the paving. The surface of the concrete can be finished in a variety of ways; see page 78 for details.

sand
concrete

HOW TO LAY BRICKS IN SAND

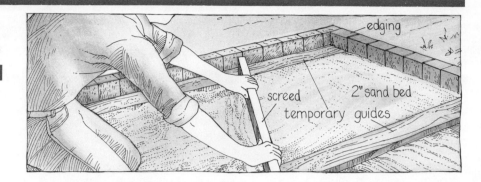

After grading the area to be paved and constructing edgings (see pages 58 and 59), you can begin laying bricks. These instructions apply to all kinds of concrete pavers, too.

1 Screed the base. Set temporary guides inside the edgings, their top surfaces one brick-thickness below the finished grade. If you use 2 by 4s, as shown, the sand bed will be approximately 2 inches deep. Place dampened sand between guides; screed it smooth, about 3 feet at a time, with a straight piece of lumber. Tamp the sand, then rescreed as necessary. (An alternative, bladed screed is shown on page 62, top; you can use it for screeding sand, if desired.)

2 Set the bricks. Working from a corner outward, place the bricks, tapping them into place with a mallet or piece of wood. (For brick bond patterns, see page 56; to see how to cut a brick, see page 39.) A mason's line (shown) aids alignment. Remove temporary guides as you work, and use a trowel to fill in the sand.

3 Sand the joints. Spread fine sand over the surface of the finished paving. Let it dry thoroughly, then sweep it into the joints, resanding as necessary to fill them. Finally, use a fine spray to wet the finished paving down; this helps settle the sand.

OTHER MASONRY UNITS IN SAND

You can lay adobe, stone, and tile in sand using the procedure shown above. There are a few differences, however, and these are outlined below.

Adobe. Sand-bedding is best for adobe. Use the same 2-inch bed you'd use for brick, but take extra care that the blocks don't straddle humps or bridge hollows; otherwise, they may crack.

Leave 1-inch open joints between blocks; this makes up for

irregularities in block sizes and allows you to pack the joints with sand or earth and crevice planting. Running bond, jack-on-jack, and basketweave patterns all work well (see page 56). The latter two do not require cutting. If you need to cut a

block, however, it's easy to do with hammer and brickset (see page 39), or with an old saw.

Stone. Lay stones in a 2-inch sand bed, following the directions for brick (above). Scoop out or fill in

sand as necessary to compensate for variations in stone thickness. If you're using irregularly shaped stones, lay them out in advance, adjusting pattern and joint spacing as needed. Use a level as you work. Permanent edgings are optional.

Tile. Heavy, ¾-inch-thick tile can be laid in sand. Follow the directions for

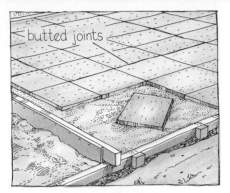

brick (page 60) but use a ½-inch sand base; a thicker base may allow the tiles to tilt out of position. Use butted joints (tiles laid with edges touching), if possible, as these will give a little added stability.

wooden tamper

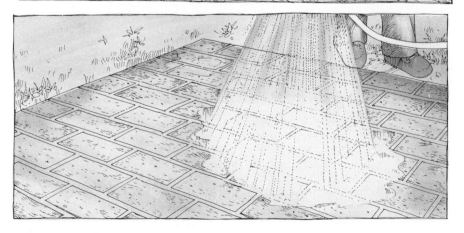

HOW TO LAY MASONRY UNITS IN DRY MORTAR

For dry-mortared brick, stone, and tile pavings, follow steps 1 and 2 on the facing page, then the steps below. Don't use this method for adobe or interlocking concrete pavers (see page 62).

1 Place bricks and mortar. Set the units with ½-inch open joints (use a ½-inch thick wooden spacer and a mason's line for alignment). Mix dry cement and sand in a 1:4 ratio and spread over the surface, brushing it into the open joints. Kneel on a board to avoid disturbing the paving. Carefully sweep the mortar off the unit surfaces.

2 Tamp the mortar. Use a piece of ½-inch-thick wood to tamp the dry mix firmly into the joints; this improves the bond. Carefully sweep and dust the unit faces before going on to the next step; any mix that remains may leave stains. (Some staining is usually unavoidable with this method.)

3 Wet the surface. Using an extremely fine spray so as not to splash mortar out of the joints, wet down the paving. Don't allow pools to form, and don't wash away any of the mortar. Over the next 2 to 3 hours, wet the paving periodically, keeping it damp. Tool the joints when the mortar is firm enough (see page 42). After a few hours, you can scrub the unit faces with a burlap sack to help remove mortar stains. For further clean-up, see page 35.

HOW TO LAY BRICKS IN WET MORTAR

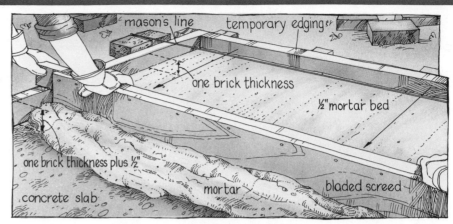

Bricks wet-mortared over a concrete slab make a firm, secure paving. The slab can be an old one, or one you've just made (see page 74). Wet down your bricks several hours before you start work; this will prevent them from absorbing too much water from the mortar.

1 **The mortar bed.** Place and screed a ½-inch-thick wet mortar bed between edgings set against the slab (for edgings, see page 59). Use a 1:4 cement-sand mix. (For more information on mortar, see page 33.) The drawing (above right) shows a bladed screed that rides on the edgings and extends down one brick-thickness below them; the edgings are set for this depth plus ½ inch, to allow for the mortar bed. Mix only as much mortar as you can use in an hour or so, and screed only about 10 square feet at a time.

2 **The bricks.** Place the bricks in your chosen pattern (see page 56), leaving ½-inch open joints between them (use a wooden spacer). Gently tap each one to bed it. Use a mason's line and a level for alignment.

3 **The joints.** Use a small trowel to pack mortar (same mix as the bed) into the joints, working carefully to minimize spilling. Tool the joints with a concave jointer, broom handle, or other convex object (see page 42). Scrub the paving several hours later with a burlap sack to remove mortar "tags" and stains. For further cleaning, see page 35.

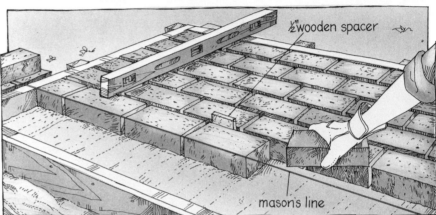

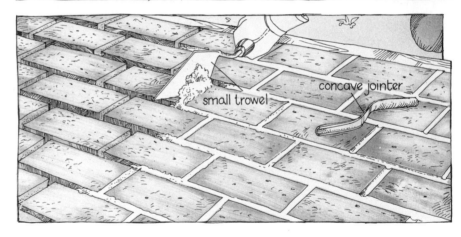

OTHER MASONRY UNITS IN WET MORTAR

You can use the wet-mortar method to set other masonry units—the best candidates are tile and stone. Don't use it with adobe and interlocking concrete pavers.

Adobe should always be set in sand, with large, open joints (see page 60). It's not advisable to use mortar. Adobe is sturdy stuff (an adobe-in-sand paving at Sunset's home office has survived more than 1.5 million visitors in the last 25 years), but salt action that occurs at mortar joints can cause block edges to crumble.

You can't use mortar with interlocking concrete pavers because there's no room for mortar in their tight-fitting joints.

To wet-mortar tile and stone, read the directions for bricks, above, then go on to the discussions below and on the next page.

Tile and stone are handled a bit differently from brick. Tile should be set in a 1-inch mortar bed for extra support, and you should use less water in the mortar than you would for brick because even the most porous tile is less absorbent than brick.

Stone should also be set in rather stiff mortar—again, because it is relatively non-absorbent. Unless you are using sawn stone, such as slate, you'll need to place mortar on the slab stone by stone, varying the thickness of the setting bed to make up for variations in stone thickness.

Tile in wet mortar. The drawing below shows tile laid in a 1-inch mortar bed over a concrete slab, using the same basic method as for brick (see facing page). The edgings shown can be left in place or removed after the mortar has set. The drawing shows a bladed screed that rides on the edgings and levels a 1-inch mortar bed; when tiles are laid on top, they will be flush with the tops of the edgings.

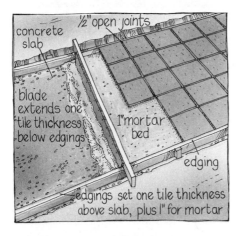

Wait 24 hours after setting the tiles, then grout them with a 1:3 cement-sand mortar (see page 34 for mixing instructions). The mortar should be just thin enough to pour. Use a bent coffee can to fill the joints, cleaning away spills immediately. Tool the

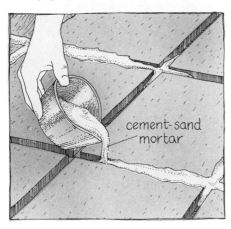

cement-sand mortar

joints as you would for brick (see page 42), using a pipe or convex jointer.

Tile can also be laid over a wood floor, but be sure to check with your supplier to find out how much extra weight will be involved; then consult your building department to see if your floor can carry it.

The drawing below details the layered construction, but see your supplier or a tile setter for more information about your particular situation. You'll also find full instructions in the *Sunset* book, *Remodeling with Tile*.

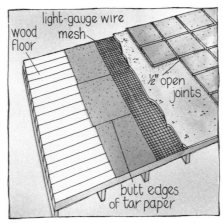

light-gauge wire mesh

wood floor

½" open joints

butt edges of tar paper

Stone in wet mortar. Arrange stones in pleasing pattern, allowing minimum space for mortar joints. To trim a stone, lap it over its neighbor and mark the trim line (see drawing below). Cut the stone by scoring the

trim line

concrete slab

line with a brickset or stonemason's chisel, then propping the edge to be cut off on a piece of wood and striking the scored line with brickset (or stonemason's chisel) and hammer.

To set stones, trowel enough 1:3 cement-sand mortar onto the slab to make full mortar beds for one or two stones at a time. Keep the mortar stiff enough to support the stones (for more on mortar, see page 33). Stones should be clean and dry to ensure a good bond.

Set each stone firmly in place and bed it by tapping with a rubber mallet. Use a straightedge and level to maintain an even surface.

rubber mallet

Let the mortar set for 24 hours, then pack mortar between the stones. Add ½ part fireclay to the mortar if desired, to improve workability. Do not use lime; it may leave stains. Smooth joints with a trowel, and clean up spills with a sponge and water. This is important, as muriatic acid washes, commonly used to clean cured masonry, can be harmful to stone.

STURDY WALLS

Poured concrete is the modern age's answer to stone. It is integral to almost all foundation work, and no high-rise structure could be built without it. Concrete has made possible everything from fencepost anchors to freeways.

If you work with masonry, you'll surely need to work with concrete, if only to get your project — literally — off the ground. A stout concrete footing is basic to all masonry walls, whether they are built of units (see pages 36–55) or poured in place. By itself, concrete can also create attractive monolithic walls and durable paving.

This chapter will show you how to construct a poured-concrete footing, how to cast a wall in place upon it, and how to pave with concrete.

Poured concrete, pro & con

Poured concrete's long suit is its strength, especially when reinforced with steel (see page 68). It usually takes considerable planning, preparation, time, and effort, though, to harness this strength.

Concrete is a mixture of Portland cement, sand, aggregate, and water. Cement is the "glue" that binds everything together and gives the finished product its hardness. The sand and aggregate (usually gravel) act as fillers and control shrinkage.

Hydration is the key to hardness in finished concrete. Hydration is a process whereby cement and water combine chemically. Some care must be taken to ensure that this process proceeds slowly and completely; if water is allowed to evaporate too quickly from concrete, hydration will be incomplete, and the finished concrete will be weak.

To ensure that hydration will be complete, wet-curing is used. This is accomplished with a concrete slab by covering the freshly placed slab with plastic or a spray-on film (available from your masonry supplier), or simply by keeping it wet for several days. Concrete walls are easier to deal with — the wooden forms in which they are made usually provide adequate protection against fast drying. (For more on curing, see pages 69 and 75.)

Poured concrete is not a casual material — a patio or a low concrete wall will probably involve more time getting ready for the pour than actually pouring and finishing the concrete.

To prepare for a concrete pour, you must build and mount a form (see page 67). For a patio slab or foundation, the form may consist of nothing more than boards firmly staked in place. The form for a large wall, though, will require considerable lumber, and the expense may not be justified for one-time use on a do-it-yourself project. Commercial contractors can re-use their formwork on dozens of jobs; you probably won't have that option.

If the scale of your project is small enough, you can mix the concrete with a hoe in a wheelbarrow (see "Mixing concrete," page 66). For a large wall or patio, it's probably best to call in a concrete truck; this will complicate the project but will save you much labor.

The appearance of poured concrete is limited only by your ingenuity. Early modern architects often took a "warts and all" approach to its texture, reveling in the raw look left by rough construction forms.

If such a raw finish doesn't suit you, you can build your forms of lumber with a pronounced attractive grain. Or you can get a good finished effect on formed concrete by whisk-brooming it with a soupy mixture of three parts fine sand and one part cement. You can also wash or sandblast a wall or paving to expose the concrete aggregate, or embed rocks and stones in it — the possibilities go on and on (see pages 78–79).

HOW TO WORK WITH CONCRETE

Once you've decided to go ahead, you'll need to make a series of decisions. The following discussion will help to guide you through these decisions; as you read, you can choose the options that best fit your project.

Buying concrete

Here you have several choices to make. Depending upon how much personal involvement you want, you can make up your own mix from scratch, buy it ready-to-mix, or order it ready-to-go.

Bulk dry materials. You usually save money by ordering your materials and doing your own mixing. For small projects, though, surcharges for small-quantity delivery can eat up your savings, so check carefully before ordering, and explore the alternatives. The most economical method is to haul the materials yourself.

Dry ready-mix. Bagged, dry, ready-mix concrete is hard to beat for convenience. Though it is a very expensive way of buying concrete, it can be the most economical on small jobs. The standard 90-pound bag, containing a mix of sand, cement, and aggregate, makes ⅔ cubic foot of concrete — about enough to set a fencepost. You can also buy bagged sand-gravel mixes. With these, you add cement and water.

Ready-mix. Some dealers supply trailers containing wet, ready-mix concrete. These carry about 1 cubic yard of concrete; you haul it yourself with your car. The trailer may have a revolving drum to mix the concrete as you go, or it may be a simple metal box into which the plastic concrete is placed. A word of caution: these trailers are very heavy;

be sure your tires and brakes are in good shape and that your vehicle is rated for the haul.

Transit-mix. A commercial transit-mix truck is the best choice for large-scale work. The truck can deliver a large quantity all at once, so you can finish big projects in a single pour. Locate concrete plants in the Yellow Pages; many have minimum orders, so be sure to check.

Special pumping equipment can be supplied to reach awkward spots. The pump forces concrete through a large hose that can be run over fences and around houses.

Choosing a concrete formula

For most residential wall and paving projects, the concrete formula that follows will give good results. You'll need to choose between the basic mix and one containing an air-entraining agent.

Basic concrete. Use this formula for regular concrete. All proportions are by volume.

    1 part cement
    2½ parts sand
    2¾ parts stone and gravel
        aggregate
    ½ part water

The sand should be clean river sand (never use beach sand); the gravel should range from quite small to about ¾ inch in size. The water should be drinkable—neither excessively alkaline nor acidic, nor containing organic matter.

If you use a shovel to measure the dry ingredients, figure about 3 quarts of water for each 6 to 7 shovelfuls of cement.

Air-entrained concrete. Adding an air-entraining agent to the concrete

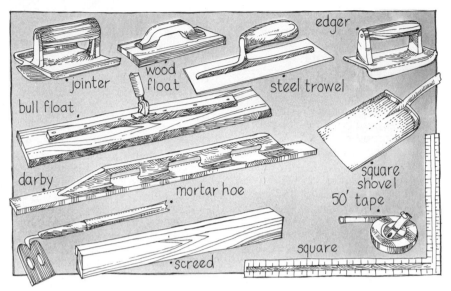

A tool kit for concrete may be already lurking in your tool shed or workshop. Shown are specialized mason's tools, but only the square trowel, jointer and edger really need to be purchased; garden tools will do, and floats can be homemade.

mix creates billions of tiny air bubbles in the finished concrete. These help it to expand and contract without cracking, a quality important in areas with severe freeze-thaw cycles. The agent makes concrete more workable and easy to place—the extra workability means you can add less water to a batch. Within limits, this makes the finished concrete stronger. Air-entraining agent is routinely added to transit mix, whatever the local climate.

The amount of agent you'll need to add will vary by brand, so consult your supplier. He or she can also tell you about any other adjustments to the basic concrete formula that may be necessary.

How much to buy. Refer to the table below to estimate what you'll need. It's a good idea to figure about 10 percent extra—you'll waste some, and you don't want to run short. You can always use any leftovers for

steppingstones or other small projects.

Here's what you'll need for each 10 cubic feet of finished concrete:

Separate ingredients:

Cement	2.4 sacks
Sand	5.2 cubic feet
Gravel	7.2 cubic feet
Ready-mix	.37 cubic yards

To use the table, figure the cubic feet of concrete needed for your project, then round the figure up to the nearest 10 cubic feet and divide by ten. Multiply this new figure by the numbers on the table to find how much of each ingredient you'll need. Round off any fractions to the next whole unit higher. If you order bulk materials sold by the cubic yard, remember that each cubic yard contains 27 cubic feet. Dry materials also are sold by the cubic-foot sack.

OF CONCRETE

Choosing a method of delivery

Your choice of delivery method should be based on the possible speed of the pour, which in turn is based partly on the size of the job and partly on the weather. If possible, the pour should be done all at once. If it is to be done in stages, plan to complete separate sections in single pours. (A single pour may require several batches of wet concrete, but no curing or drying should occur between batches.) Never interrupt a pour once it has begun, and remember that hot, dry weather will substantially shorten your available working time.

Forms can be filled from a wheelbarrow—you can either dump (you might need a ramp) or shovel. You'll find a splashboard will save concrete by putting it where you want it (see drawing on facing page). Or you can pour the concrete directly from the drum of a power mixer placed next to the form.

Unless your job is very small, you may need some help when it comes to the actual pour. If you're doing your own mixing, you'll find it helpful to have one person mixing while others wheel and place the concrete. If you call for a transit-mix truck, you'll certainly need extra hands, especially if you have to move the concrete from one place to another.

Mixing concrete

If you decide to mix your own concrete, you have two choices: hand mixing or power mixing. For small projects, hand mixing is undoubtedly the simplest method. Large forms that must be filled in a single pour, though, may warrant a power mixer. Generally, a mixer smaller than 3 cubic feet is more nuisance than it's worth, and a mortar box is a better bet. A small mortar box (53 by 25 by 11 inches) will readily take up to 6 cubic feet of concrete—a sack of cement at a time.

If you are using air-entrained concrete, you'll have to choose a power mixer. Hand mixing is simply not vigorous enough to create the air bubbles in the concrete mix. Whatever the method, add water in small amounts; too much can ruin the mix.

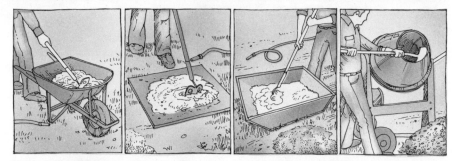

Choose a method of mixing, based on your needs. A high-sided wheelbarrow is good when you need to move the concrete; if you don't need to move it, a simple sheet of plywood may do. For bigger jobs, a mortar box is good; really large jobs may require a power mixer.

Hand mixing. A high-sided contractor's wheelbarrow is okay for mixing if you don't need to work more than 1 to 2 cubic feet at a time. You can mix larger batches on a simple wooden platform or in a mortar box, either one homemade of plywood or a ready-made model available at masonry supply stores. Plan to work in small batches; this will make mixing easier and give you greater control over proportions.

Ingredients for small quantities of concrete are usually measured with a shovel—it's plenty accurate if your scoopfuls are consistent. If you need greater accuracy you can use a 1-cubic-foot wooden box (make it bottomless for easy dumping—lifting the box empties the contents). Or empty a sack of cement into a bucket; level, and mark the bucket; this equals 1 cubic foot. If you also mark the bucket for ½ cubic foot, you'll find it easier to lift.

Mark a bucket off in quarts and gallons to keep track of the water. And set up so that you can bail water from a drum or your garbage can; it's more convenient than turning a hose on and off.

Here's how to mix a small batch according to the proportions given in "How much to buy" (page 65): Place 2½ shovelfuls of sand on the mixing surface and add 1 shovelful of cement, mixing thoroughly; add 3 shovelfuls of gravel and mix again; mound up the mixture and hollow out the center; using the marked bucket, pour in ½ quart of water; work around the hollow, pulling the dry ingredients into the water, always enlarging the size of the basin. Continue mixing until the concrete is all the same color and all dry ingredients are thoroughly wet. This makes a small trial batch, allowing

you to check and adjust proportions (see "The trial batch," below).

Machine mixing. You can rent, borrow, or buy cement mixers in sizes ranging from ½ to 6 cubic-foot capacity, but the smaller ones (under 3 feet) don't pay off. A mixer can be either electric or gas-powered. Set the mixer close to your sand and gravel piles so that you can shovel-feed direct; be sure it's level, and chock it in place to prevent "walking."

To mix ingredients, start the mixer (warm it up if it's a gas one) and add a little water.

Then, and in order, add a little gravel and sand, more water, more gravel and sand, and last the cement. (Measure your ingredients by shovelfuls as you add them.) Caution: Don't put the shovel inside the mixer—you could get your skull cracked. Check the mix by pouring a little out—never look inside a mixer that is running. Mix for a few minutes—just enough to get everything worked in and all the particles wet.

The trial batch. Work a sample of your first batch with a trowel. The concrete should slide—not run—freely off the trowel. You should be able to smooth the surface fairly easily, so that the large aggregates are submerged. Both large and small aggregates at the edges of the sample should be completely and evenly coated with cement.

If your mix is too stiff and crumbly, add a little water. If it's too wet and soupy, add sand-cement mixture. Be sure the sand and cement are correctly proportioned according to your recipe. If you do make adjustments, be sure to record them accurately; don't rely on "feel."

BUILDING CONCRETE FOOTINGS & WALLS

No matter what type of masonry material you are using for your wall, concrete is the material of choice for the footing. The discussion that follows will explain the details of form building so that you can form up the footing. It will also detail how to cast a poured concrete wall on top of it.

Many municipalities restrict freestanding masonry walls over 3 feet high; retaining walls are also usually tightly controlled. You'll need a building permit if you go higher, and you may need to have the wall designed by an engineer. For these reasons, the discussion and drawings that follow describe a freestanding wall 3 feet high or lower. Check your local building department if you plan to go over this limit. Retaining walls are described on page 88.

FORMS FOR FOOTINGS & WALLS

Designing and building forms—the wooden structures which hold and mold the wet concrete to create your footing or wall—is not difficult, but it is exacting. Concrete is dense and heavy; it exerts a lot of pressure on the forms, so they must be very sturdy. The drawings that follow show several standard methods of building forms. There are many others. One of the intriguing aspects of form design lies in coming up with interesting embossed patterns by using rough-grained lumber or alternating board sizes.

From the ground up: The footing

Typical footings are twice the width of the wall and equal in depth to the wall's width (see page 37), but be sure to consult local building codes for exceptions to this rule of thumb.

All footings should be laid on a gravel bed, usually about 6 inches thick, that lies beneath the frost line. The underlying earth should be firm, ideally of uniform consistency.

The easiest route to a footing is simply to dig a trench and pour concrete in place in the earth. In cases where the earth is too soft or too damp to hold a vertical edge, you can build a simple form, as shown in the drawing below.

The footing should be flat on top if a unit masonry wall (see pages 36–55) is to be built on top of it. If you plan to cast a concrete wall in place on top, you have two choices. For lower, lighter walls, you can cast the wall at the same time you cast the footing (see drawing below).

Larger walls will require a separate pour in a sturdy form and should be

keyed into the cured footing as shown below.

Formwork for concrete walls

If you understand the principles of formwork for a straight wall, you will be able to handle the formwork requirements of your specific project. Above all, your form must be strong. It will have to withstand the pressure of wet concrete, plus the jostling it will get during the pour.

A straight wall form (see drawing below) is made up of sheathing, studs, spreaders, ties, and (for larger walls) wales. Sheathing forms the mold. Studs back up and support the sheathing. Spacers set and maintain spacing and also prevent collapse of the form prior to the pour. Ties snug up the form and resist the pressure of the wet concrete. Wales align the form and brace the studs in larger forms. A splashboard helps direct the concrete.

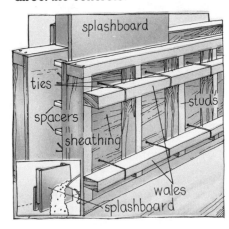

Building the form. Forms for straight walls are built in sections. First, with the lumber flat on the ground, make a frame of 2 by 4s on edge; it should be at least as high as your completed wall, and 8 feet is a workable overall length. Next, 2 by 4 studs are nailed in; 16 to 19-inch centers are adequate for most do-it-yourself projects, but space them closer if you plan an especially thick wall to be poured all at once.

The frame is now ready for sheathing; use either ½ or ¾-inch plywood or 1-inch lumber.

Next, lay your sections down, sheathing sides face to face, so that you can position the wales, if any, and drill holes for the ties. Wales are horizontal 2 by 4 braces that also align your form and anchor the ties. Two wales are enough for most jobs, but don't exceed 30-inch spacing. Plan for your wales to be about 6 to 8 inches above the form base and below the form top. Mark and drill ⅛-inch tie holes through the sheathing above and below each wale midway between studs. Now, toenail the wales in place (a few spots will do). Wales can be made of several pieces if you overlap them a foot or more; each piece must bear on at least two studs.

You will need spacers and ties to assemble sections and hold the proper space between them. Make spacers of 1 by 2s or 2 by 4s; cut them the same length as the finished thickness of your wall, and make enough to put them in about every 2 feet, horizontally and vertically. There are commercial ties that the pros use, but you'll probably find wire ties (8 or 9-gauge iron wire) your best bet. Cut the ties long enough to encircle opposite wales, plus a couple of inches for twisting the ends.

Tilt two sections upright, face to face, and spaced at wall thickness; tack the tops with several crosspieces to hold the panels in place. Thread a tie through the form and around the wales, and twist the wire ends together. Place a spacer near the tie.

Now, put a stick between the two wires inside the form and twist them as shown until they are tight. Remove stick and repeat until all spacers and ties are in place. You will have to remove spacers as the

pour is made, so tie long pull wires to any of them that will be out of reach.

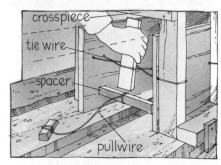

Add form sections to this initial one to make up the entire wall form; butt the sections and nail through the adjacent 2 by 4 studs. The running length of the side panels should be several inches greater than the finished wall so that you can cleat in stop boards at the ends of your form (see drawing below).

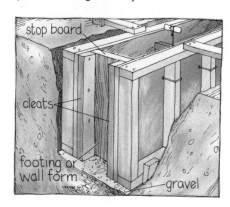

Pouring in stages is a good idea if the wall is very long. In this case, section the wall vertically and use a stop board, secured by cleats, between pours. Allow the concrete to set up for a day or two, then move stop board and make the next pour. Steel reinforcing, which must always be continuous, can be run through holes in the stop board.

You may not need the full treatment just described. On low walls, you can get by with simpler forms. Build only to your need.

Curved walls. You can make curved forms by saw kerfing either plywood or lumber and bending it toward the kerfs (see page 73). Cut about two-thirds through the wood using a power circular saw. Soak the wood well before bending it, to make the job easier. The curved form does not use wales.

Bracing the form. Properly constructed, a form is self-supporting.

But the form must be plumbed and tied in place with braces — it has to stay put during the pour. To prevent wet concrete from lifting the form, nail the bottom to stakes driven in the ground. Lateral movement is prevented by 2 by 4 angle braces, as shown below.

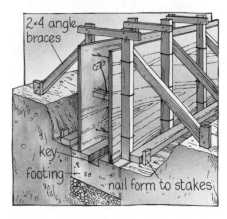

Preparing the form. Just before pouring the concrete, coat the form with a commercial release agent or motor oil. This will make it easier to remove the form.

Steel reinforcing for maximum strength

Like other masonry materials, concrete is enormously strong in compression. A concrete column can resist crushing forces of thousands of pounds per square inch. But if a rod or chain were made of concrete, it would snap before it could lift a fraction of the weight it supported in compression. The force that snaps it is tension.

Steel changes all this; placing steel rods where tension forces concentrate enables concrete to resist the tension, making it unique among masonry materials.

In a retaining wall, the steel runs up the back, where it can best resist the toppling force of the earth that tries to bend it forward. A freestanding wall that takes wind forces — and perhaps an occasional automobile — from both sides is reinforced in the center, as a compromise.

Local codes specify exactly how much and what kind of reinforcing you'll need; for this reason, we cannot specify details here. Building to code is especially important for retaining walls and large foundations, which should be professionally designed for real security.

guide line

batterboard

sand

HOW TO POUR A CONCRETE FOOTING

Every wall needs a stout concrete footing. Here's how to make it.

1 Lay out the footing. Use strings and batterboards as a guide for the footing trench. You can mark a line on the ground if you lay a taut string, cover it with a small bit of sand, and then lift the string. You also use strings as guides for your forms once the trench is complete.

tamper

gravel

2 Prepare the base. Level the bottom of the trench and tamp it firm, using a rented or homemade tamper. On sloping sites, plan to use a stepped foundation (see step 3 below). Place a 6-inch gravel base in the bottom of the trench. The top of the gravel should lie below frostline for your area.

footing for unit masonry

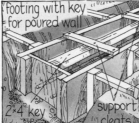

footing with key for poured wall

2"x4" key

support cleats

stepped footing

tie joints with wire

reinforcing

3 Construct form. To build your form, follow the directions on page 67, using the batterboard strings as a guide. Three variations are shown. Steel reinforcing can be supported on pieces of brick, stone, or broken concrete. It should be placed ⅓ of the foundation's thickness up from the bottom. If needed, a 2 by 4 supported by cleats can be added to form a key in the footing. Do this if you plan to cast a concrete wall on top.

4 Pour and tamp the concrete. Oil the form with motor oil and pour the footing, tamping it firmly with a shovel. Run the shovel up and down along the edges of the forms to ensure that no voids are left. In pouring and tamping, always work systematically from one end of the footing to the other. At the far end, stop the pour short of the top to allow for screeding.

screed

2"x4" key

remove support cleats after pouring

5 Strike off the forms. Use a piece of wood to "strike" or screed the concrete level with the top of the form. Work zigzag from one end to the other, knocking down high spots and filling any hollows. If you are adding a wall that requires vertical reinforcing bars, insert them at this time. Remove the key's 2 by 4 as soon as the concrete will hold its shape.

6 Cure the concrete. After striking, the footing should be cured for 4 days. The slower the set, the stronger the concrete. Cover it with plastic sheeting, burlap, or newspapers. Keep porous coverings wet: spray them several times a day with a hose, especially in hot, dry weather. Once curing is complete, remove the forms.

burlap

finished footing with key

Concrete Footings

HOW TO CAST A CONCRETE WALL

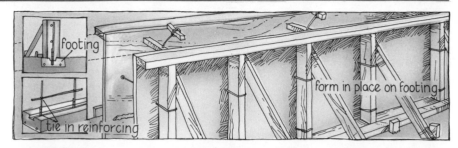

With the footing in place, you can cast a wall on top. Here's how.

1 Mount forms on cured footing. See page 67 for details of form building. If you are using reinforcing bars, tie them into the foundation reinforcing, as shown at right. Generously overlap them and tie together in several places with wire.

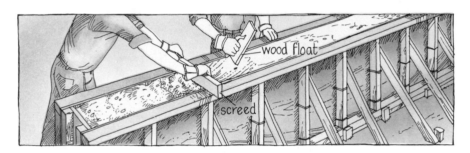

2 Pour and tamp the concrete. Follow the same procedure you used to pour the footing (see page 69), but take extra care. If you want a smooth finish, be sure to work the concrete well up against the sides of the form. Pour and tamp concrete in 6 to 8-inch layers, pulling out the spacers as you go, and working the concrete in and around the reinforcing bars.

3 Strike, float, and trowel. Strike off the top of the forms as you did for the footing. If you want a smoother finish, follow with a wood float. The smoothest finish is obtained with a steel trowel; see the section on concrete paving, page 75, for details on its use. Cover and cure the wall for 4 days , as you did the footing, before removing the forms, unless you intend to texture the wall surface (see below).

4 Two alternative caps. When you've filled the forms close to the top, you can set anchor bolts in the concrete to hold a wooden cap in place after the forms are removed (left). At right, a small form is mounted on the cured wall, the top of which was left rough after striking. The form is used to cast a flared cap on the wall after removing the wall forms, as shown. Quarter-round moldings added to the form give a decorative effect.

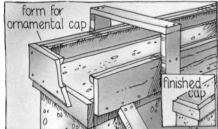

TEXTURED SURFACES FOR CONCRETE WALLS

If you're not satisfied with the "plain vanilla" look of a smooth, cast-concrete wall, you can achieve a more textured surface by using either of the methods outlined here. In the first, the popular exposed-aggregate paving finish is adapted to walls; in the second, stones and concrete combine for a rough-hewn effect.

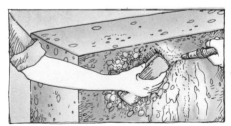

Exposed-aggregate surface. You can create attractive exposed-aggregate textures by first spraying the inside of the forms with a special retardant that delays the hardening of the outside layer of concrete. The forms are removed early and the surface is scrubbed and hosed, or sandblasted (not usually a do-it-yourself technique), to reveal the aggregate. The time of form-stripping is critical, so be sure to consult an expert.

Stone-in-concrete. This technique—pioneered by Frank Lloyd Wright—produces a rugged surface and saves on concrete. Selected clean stones are placed against the forms as the pour proceeds; their faces are revealed when the forms are removed. Sometimes it's necessary to chisel away some of the concrete from the faces of the stones.

PAVING WITH POURED CONCRETE

When it comes to pavings in and around the house, you'd be hard put to find a more adaptable material than poured concrete. From a few simple ingredients—cement, sand, gravel, and water—you get a hard, durable surface that can be shaped to fit almost any situation.

With it, you can create anything from small steppingstones to sizable slabs. Its surface can be made smooth enough for shuffleboard, or rough enough to keep you from slipping on a steep pathway. Concrete can be colored, embedded with attractive stones, or made to resemble stone itself. Concrete is what you make it—and it's always practical and serviceable.

In this section you'll learn how to pave with concrete—how to build forms, pour, and finish the concrete. For information on ordering and mixing concrete, refer to pages 64–66.

Before you begin: A few cautions

Laying your own concrete paving can save you money, compared with the cost of a professional job, but you'll need to be realistic and plan ahead.

Divide your work into stages that you and one or two other people can handle. Pour large areas in sections, cast a few paving stones at a time—this way you can compensate for the large work crew and specialized equipment the contractor has at his disposal.

As always with building projects around the home, check with your building department about code restrictions that may apply to your project. A garden path will likely be no problem, but a major patio project might. You'll need to take into account the soil, drainage, and frost conditions for your area. Local officials and engineers can be a real help here.

Once concrete is in place, you're stuck with it; repair and replacement are costly and much more difficult than doing the job carefully and correctly in the first place. Take your time in planning and preparation, and be sure the site is ready.

PREPARING FOR THE POUR

Like unit paving, concrete requires a stable, well-drained base. Because the finished slab is monolithic, it is especially important to ensure that the ground beneath it doesn't shift and cause the concrete to crack.

Preparing the base

Refer to the chapter on unit paving (pages 56–65) for directions on laying out and grading the site for your paving. Plan on at least a 2-inch gravel base in areas where frost and drainage are not problems, and a 4 to 6-inch base where they are. (The shallower base can be made of sand if drainage is especially good; gravel is best for deep bases.)

Using a square-ended shovel, cut and remove turf to the required depth (slab thickness plus gravel thickness); make the trench about a foot wider and longer than the finished slab is to be. If the exposed earth is soft, wet it and compact it firmly with a commercial or homemade tamper.

tamper

strings align trench
batterboards

Once the excavation is complete, build and place your forms (see below), then add the gravel. It's easier to level the gravel base if the forms are already in place, because you can use them as a guide for a special bladed screed, as shown in the drawing below. Allow the gravel to extend under the form to the edges of the excavation.

thickness of finished slab

gravel

bladed screed

Building & placing the forms

The standard slab for pathways and patios is nominally 4 inches thick. Usually 2 by 4s are used for forms in concrete paving work, and actual thickness will be 3½ inches if you use finished lumber. If you want a thicker slab (or a more rustic appearance), you might try rough, undressed lumber, which has slightly larger dimensions.

If you plan to leave the form lumber in place as permanent edgings and dividers, be sure to use redwood or cedar. Get heartwood if you can, and be sure to treat the wood with a preservative before committing it to the earth. Pressure-treated lumber is also a good choice.

Straight forms. These are the easiest to build. They can be assembled

A NEW WRINKLE IN CONCRETE — STAMPING

The look of unit masonry with the rapid coverage and strength of poured concrete — this is an attractive combination for commercial paving contractors and their clients. Their method of achieving it — concrete stamping — is finding increasing use in residential work. Though do-it-yourself concrete stamping is probably just around the corner, you have to hire a contractor to get it done now.

Concrete stamping makes a slab resemble brick, adobe, or stone. The characteristic fan-shaped pattern of European cobblestone pavings is one of the most popular patterns. Several brick patterns are available; these can be left as-is, in which case they resemble brick in sand, or the stamped "joints" can be mortared so that they resemble regular brick mortar joints.

The stamping technique is simple, at least for the contractor. First, a regular concrete slab is poured and floated smooth. (Often the slab is colored, usually by the "dusting" method — see page 79.) After floating, a special interlocking grid of patterned aluminum stamps is pressed into the slab. Workmen stand on the grids to force them into the concrete. A final going-over with a trowel fixes any blemishes, and the project is cured in the usual manner.

The advantages of concrete stamping lie in potential time and cost savings. The look approximates real unit masonry, and it gives you yet another option for such projects as driveways and patios.

---

from standard lumber, either cut to length or spliced, as shown in the drawing below.

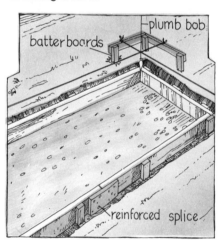

When placing the form boards, nail them to sturdy stakes driven into the ground at least every 4 feet. Use a taut string as a guideline. Drive the stakes plumb and deep enough so that they will not overlap the top of the form and obstruct the screed. Use batterboards (see drawing above) to lay out accurate corners. Nail the corners securely, and stake them well. If you plan to strip the form, use doubleheaded nails to make disassembly easier.

The wet concrete will exert considerable pressure on the form, so be sure your stakes are secure. Don't forget to space them the thickness of the form boards out from the edge of the finished slab; for 2 by 4s, this would be 1½ inches.

When nailing a form board to a stake, use galvanized nails and back up the stake with a sledge hammer, stone, or other heavy object. Drive 16-penny galvanized nails part-way into the form about every 18 inches on the inside if you plan to leave the form in place; this will lock the boards to the slab (see drawing below).

Check the level of the form from side to side by placing a bubble level on a long, straight board extending the width of the form. Your finished slab should have at least enough pitch in one direction to allow water to run off. If the natural grade of your site isn't enough, plan to pitch the slab about ¼ inch per foot. Do this by nailing one sideboard slightly higher than its mate (see drawing below).

Stepped forms. Pathways on steeper grades are much easier to climb if they are poured in a stepped form. Each extended step can have its own pitch so that the path generally follows the natural slope of the hill.

In excavating for a stepped form, you'll probably have to do some cutting and filling. Be sure to tamp thoroughly any fill you make; it is an area where settling trouble may occur later. Where possible, substitute coarse (¾-inch and larger) aggregate for earth when filling.

When you lay in the base gravel, keep it back from the actual step area (as shown in the drawing below) so that the concrete will be extra thick at this potentially weak point.

Curved forms. Curved wooden forms for slabs can be easily built in three ways: bending, laminating, and saw-kerfing.

Bending works best for gentle curves. For a temporary form, use plywood, with the outer grain running up and down, and bend it around stakes set on the inside edge of the curve. Determine the radius with a string compass, as shown.

For the outside edge of the curve, bend the plywood around temporary stakes and secure the ends by nailing them to stakes set on the outside. Add more stakes on the outside, nailing the plywood to them; then pull up the inside stakes. Using a power or handsaw, kerf the plywood (see below) if needed.

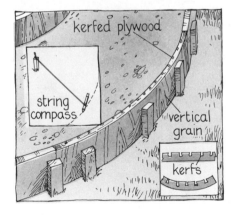

Use the laminating method if the curved form is to remain in place. Following the preceding directions,

bend several layers of redwood benderboard around the stakes until you have built up a thickness equal to your other form boards.

Really tight curves can be made only with sheet metal, plastic, or other flexible material. Cut the material to size and nail it to the stakes. You may need extra stakes to ensure adequate support.

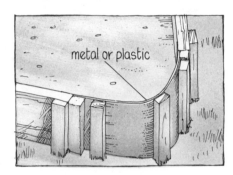

Expansion strips. If your walk or patio extends more than 20 to 25 feet in any direction, and if it doesn't have permanent wooden dividers, you'll need to put in expansion strips. These strips (available from your supplier) allow movement between adjoining sections of the slab and help to prevent cracking due to expansion and contraction of the concrete.

You can best install expansion strips by making a movable stop board, as shown in the drawing below. Before pouring the slab, set the stop board in position with stakes,

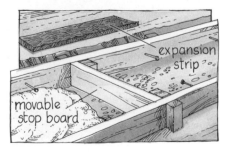

and pour the concrete up to it. When the concrete has stiffened slightly, remove the board, set in a precut length of expansion-strip material, and continue pouring.

Always plan to use an expansion strip when pouring concrete up to an existing structure, such as a house foundation. This forms an *isolation joint* that allows independent movement of the two structures.

If you're planning a walk or patio with permanent wooden forms, include wooden dividers every 8 to 10 feet or so—they will serve the same purpose as expansion strips.

Reinforcement. It's always a good idea to reinforce concrete. The steel helps to prevent cracking and will hold the pieces together if cracking does occur. If your project is more than 8 square feet, you should reinforce it with steel mesh designed for the purpose. Six-inch-square welded mesh is most commonly used.

Install the mesh after the forms are ready. Cut it to size with bolt cutters or heavy pliers, keeping it several inches away from the sides of the forms. Support the mesh on small stones, bits of brick, or broken concrete so that it will be held midway in the slab.

Preparing the forms. Just before pouring the concrete, go over your forms to check for level (or grade) and to be sure everything is secure. Temporary forms should be oiled to aid in stripping; use motor oil or a commercial release agent. Cover the top edges of permanent forms with waterproof masking tape to prevent their being stained by the concrete.

HOW TO POUR A PAVEMENT

The forms are set, and you're ready for the concrete. The following guide will take you through the basic steps of pouring, spreading, striking, finishing, and curing a concrete pavement. These steps apply whether you're making a single steppingstone (see page 76) or pouring an entire patio; only the scale is different.

Before you begin, be sure you have enough hands for the job. Except for small projects, at least two people will be needed for most concrete work; remember that once the pour begins, it should proceed right through to the final curing step without interruption.

Don't neglect your tools, either. Be sure you have enough shovels and hoes to spread and tamp the concrete. Gloves and rubber boots are a good idea. The latter are needed if you'll need to walk on the concrete to screed it; concrete is caustic, so always wear gloves to protect your hands.

1 Pouring, spreading, & tamping. Start pouring the concrete at one end of the form while a helper uses a shovel or hoe to spread it. Work the concrete up against the form and tamp it into all corners—don't simply rely on gravity. But don't overwork the concrete, and don't spread it too far; overworking will force the heavy aggregate to the bottom of slab and will bring up the "fines"—inert silt that can cause defects in the finished concrete. Instead, space out your pours along the form, working each batch just enough to completely fill the form.

2 Striking the concrete. Move a screed across the form to level the concrete. On long pours, do this batch-by-batch rather than after all the concrete is placed. Move the board slowly along the form, using a rapid zigzag, sawing motion. Even on narrow forms, two people will make the work faster and more accurate. A third person can shovel extra concrete into any hollows, as shown at right.

3 Edging & jointing. Before and after floating (see upper right), you must edge the concrete. Begin by running a trowel between the concrete and the form. Follow up with an edger. Run the tool back and forth to smooth and compact the concrete, creating a smoothly curved edge that will resist chipping. Use the edger or a special jointer with a straight guide board, as shown, to make control joints. (A control joint is a deliberate weak point where cracking can occur beneath the joint and not be seen.) Control joints should be made in long walks at intervals no greater than 1½ times the width of the walk.

run trowel along edge

follow with edger

edger

guide board

jointer

Concrete Paving

darby

bull float

wood float

steel trowel

avoid overlap

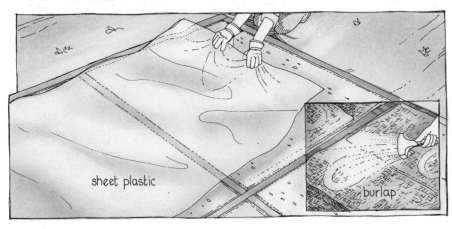

sheet plastic

burlap

4 **Floating.** After screeding and edging, use a bull float, darby, or wood float—depending upon the size of your project—for the initial finishing. Floating smooths down high spots and fills small hollows left after screeding. Push the bull float (right) away from you with its leading edge raised slightly, and pull it back nearly flat. Overlap your passes with the float. Use the darby (left) on smaller projects; move it in overlapping arcs, then repeat with overlapping straight, side-to-side strokes. Keep the tool flat, but don't let it dig in. On very small projects, a wood float can be used in a similar manner. Redo your edges and joints after floating.

5 **Final floating and troweling.** After the water sheen has disappeared from the concrete, but before the surface has become really stiff, give it a final floating with a wood float (or see "Brooming the surface," below). For a slick, smooth surface, follow with a steel trowel. Make your initial passes with the trowel flat on the surface; use some pressure but don't let the blade dig in. If you want a smoother surface, wait a few minutes and repeat the operation, this time with more pressure and with the leading edge raised. Kneel on boards to reach the center of a large slab.

6 **Brooming the surface.** If you want a nonskid surface, substitute brooming for final floating and troweling. The texture you produce will depend on the stiffness of the bristles and whether you use the broom wet or dry. The pattern can be straight or wavy, depending on how you move the broom. Drag the broom over the concrete immediately after floating, always pulling it toward you. Avoid overlapping passes; this tends to knock down the grain texture and produces too many "crumbs." Finish up by redoing your edges.

7 **Curing.** Slabs need to be moist-cured to keep their surfaces from drying too fast. If the surface dries too soon, it will be weak and may later become powdery or flake away. Cure your concrete by keeping it wet. You can do this by covering the slab with straw, burlap, or other material and wetting it down. If you cover the slab with plastic sheeting or a commercial curing compound, water evaporating from the concrete will be trapped, eliminating the need for wetting. If no covering material is available, you'll need to keep the surface damp by hand sprinkling. Curing should last at least 3 days—longer in cold weather. It's a good idea to cure your project for about a week, just to be on the safe side.

Concrete Paving

HOW TO CAST PAVING BLOCKS

Cast-concrete paving blocks are easy for everything from pathways to patios. Laid on sand with large open joints, then filled with turf or ground cover, they quickly become a part of your garden. You can buy the units precast, but why not consider making your own? The techniques are simple, and you are assured of getting something distinctive that's just right for your needs.

Three techniques are discussed here, followed by a grab bag of ideas for forms. All the techniques used in finishing and texturing concrete presented on the previous pages apply to paving blocks—review them before you begin.

Casting in ground molds. The easiest way to make steppingstones is to pour them in place in the ground.

Dig a hole 4 inches deep for each "stone," contouring it as you choose. For easy walking, space the steps no more than 18 inches apart. If you are working in a lawn, plan to keep the tops of the stones below ground level to allow for mowing.

Use a 1:2:3 cement-sand-aggregate mix to fill the holes (see page 65). You can save concrete if you toss in a few stones as filler. Finish the tops with a trowel or wood float, cover, and cure the steppingstones as described for concrete pavings on page 75. You might try the travertine texture (page 79) to enhance the look of stone.

Casting in single molds. For a more controlled edge, you can cast your paving blocks in a mold. This method lends itself to smaller, lighter blocks.

A simple closed frame is the easiest to construct. You can make it from 2 by 4 lumber or any sturdy wood. For ease in unmolding, hinge one or two corners as shown in the drawings at right and add a hook and eye to close the open corner.

To cast your blocks, oil the form with motor oil and place it on a sheet of plastic, a sand bed, or other smooth surface. Fill the box with a stiff concrete mix, packing it in, and strike off—screed—the top with a piece of wood. You can float the sur-

...IN GROUND MOLDS

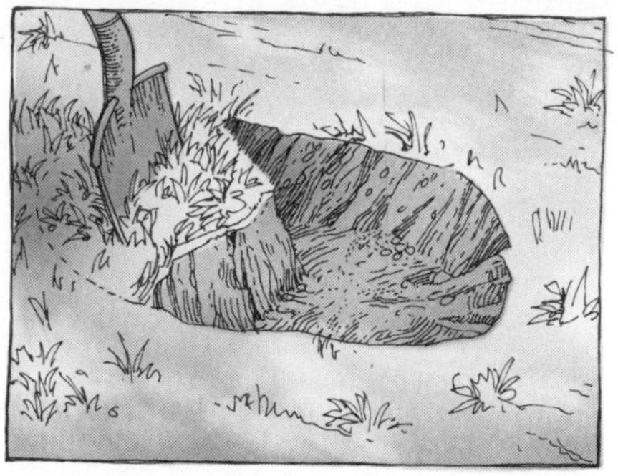

Dig a hole, keeping sides sharply cut and fairly vertical.

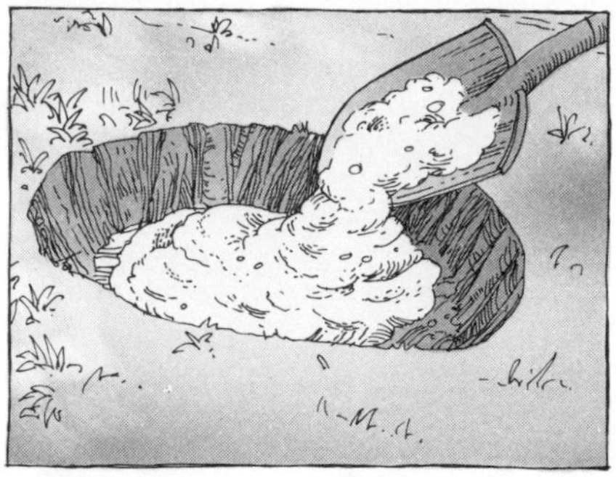

Fill the hole with concrete (see page 65 for proportions).

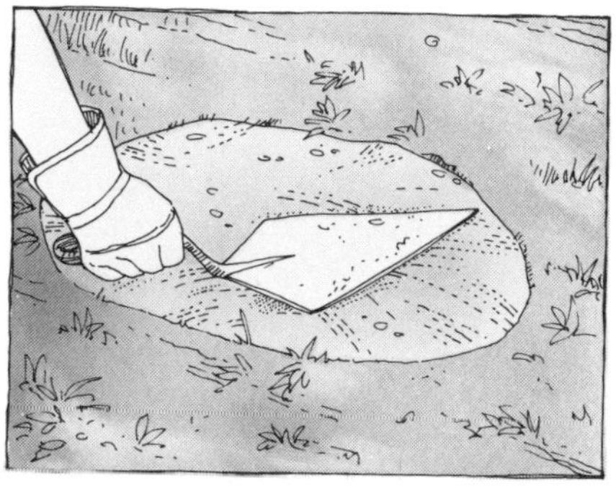

Finish the top with a trowel after thoroughly tamping the concrete.

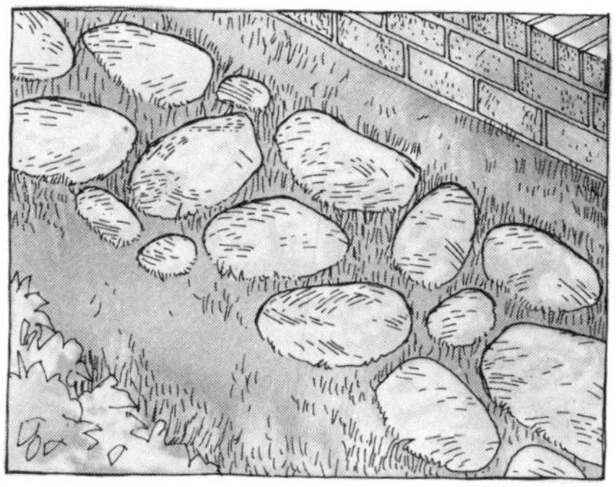

Finished pathway has an easy, natural look that resembles random stones.

...IN SINGLE MOLDS

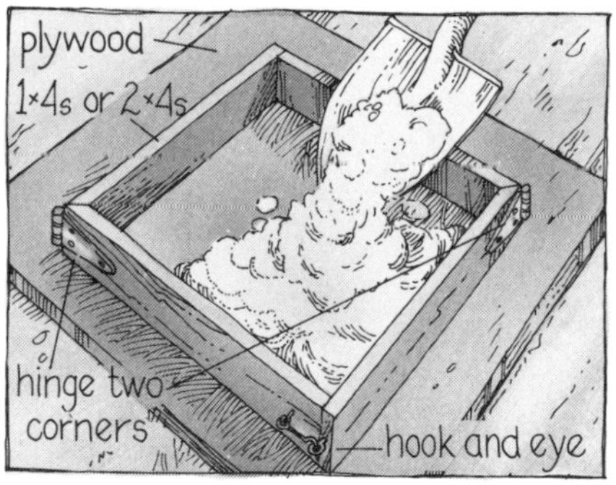

Place concrete in oiled, open mold set on oiled plywood (see left for construction).

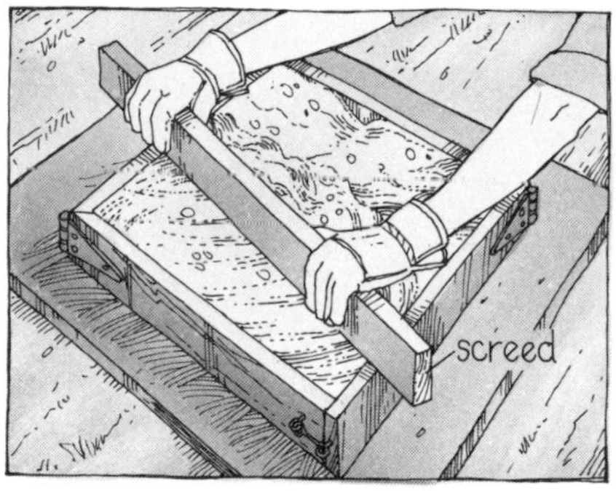

Screed concrete after tamping; use a piece of scrap wood.

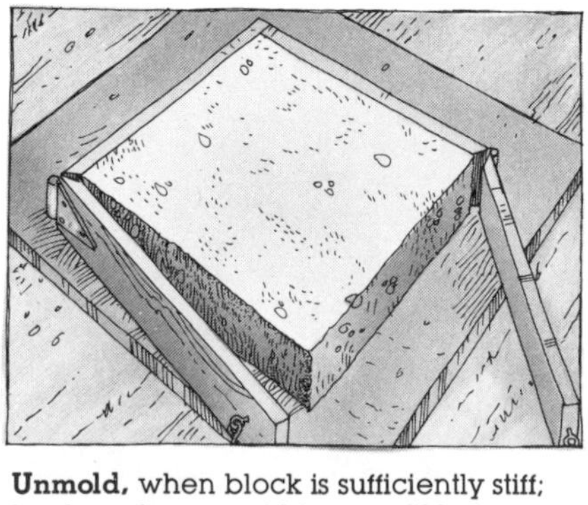

Unmold, when block is sufficiently stiff; hook and eye and hinges will help.

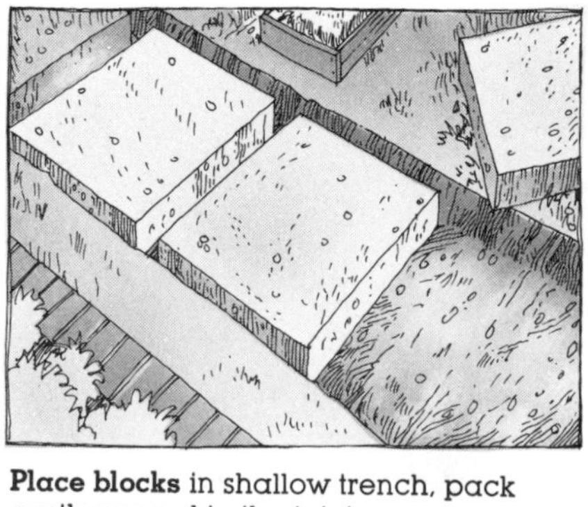

Place blocks in shallow trench, pack earth or sand in the joints.

face if desired (see page 75) or use the smooth underside of the block for the stepping surface.

Wait until the concrete has set somewhat before unmolding; this won't take more than a few hours if you've used a stiff mix, and you'll be able to go right to the next block. Cure the blocks just as you would a concrete slab (see page 75).

Placing a bottom in the box form can lead to all sorts of interesting designs, as shown in the drawings below.

Casting in a multiple-grid mold. If you have a large area to cover, such as a long walk or a patio, a grid form will speed up the job.

To cast the blocks, oil and place the form on the sand or gravel bed or graded, well-compacted soil, and fill the compartments with concrete. Strike off the form and float the concrete (see page 75). Remove the form as soon as the mix is stiff enough to hold its shape. Clean and reoil it between pours.

...IN MULTIPLE-GRID MOLDS

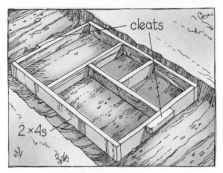

Place oiled mold in shallow trench; 2 by 4s make up the mold, cleats aid lifting.

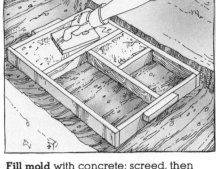

Fill mold with concrete; screed, then finish surface with trowel or wood float (shown).

Lift mold when blocks are sufficiently stiff; clean up edges with trowel.

Finish path by moving mold along and repeating steps; fill joints with earth.

A POTPOURRI OF PAVING BLOCKS

You can really let your imagination go with paving blocks. To set you thinking, a few of the many possibilities are shown at right.

All molds, especially those involving moldings and other small pieces of wood, should be liberally oiled before casting begins. It's also a good idea to press paste wax into all corners; this gives a block with rounded edges that releases easily from the mold.

Finer patterns can be brought out if you cast an inch or two of 1:3 cement-sand mix first (for more information on mortar, see page 33), then top it off with concrete. Stones or other rubble can also be placed in the mold once the tread surface is cast, to save on concrete.

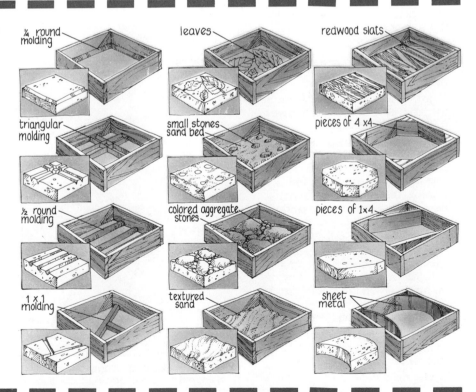

SPECIAL FINISHES FOR CONCRETE PAVING

The appearance of a concrete surface can be altered to suit a variety of purposes—and tastes. Here are some of the most widely used methods.

Exposed aggregate. The attractive exposed aggregate finish is probably the most popular for residential concrete work. There are two basic ways to produce it: "seeding" a special aggregate or large smooth pebbles into the concrete surface or exposing the regular sharp aggregate already contained in the concrete.

To expose the regular aggregate, pour and finish the concrete through the floating stage (see pages 74–75). Don't overfloat, or you may force the aggregate too deep. You'll get better results if you use a uniform, coarse aggregate.

When the concrete has hardened to the point where it will just support your weight on knee boards without denting, you can begin exposing the aggregate.

Gently brush or broom (nylon bristles are best) the concrete while wetting down the surface with a fine spray. Take care not to dislodge the aggregate, and stop when the tops of the stones show.

With the seeded aggregate method, you can use the varicolored smooth pebbles that make this finish so popular.

Pour the slab in the usual manner, but strike it off about ½ inch lower than the form boards.

Distribute the aggregate evenly in a single layer over the slab. Using a piece of wood, a float, or a darby, press the aggregate down until it lies just below the surface of the concrete. Refloat the concrete and proceed with the washing and scrubbing steps described above.

Take extra care in curing exposed aggregate surfaces; the bond to the aggregate must be strong. Any cement haze left on the stones can be later removed with a 10 percent muriatic acid solution (see page 80).

Salt finish. Coarse rock salt can be used for a distinctive pocked surface on concrete.

Scatter coarse chunks (no fines) of rock salt over the surface of the slab

SEEDED AGGREGATE

Screed concrete after placing, knocking down high spots and filling low ones.

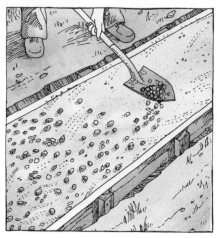

Spread selected aggregate over the surface in a single layer.

Embed aggregate with a darby or piece of wood; float surface, burying aggregate.

Brush and hose the surface gently once it is stiff enough; be careful not to dislodge stones.

SALT FINISH

Sprinkle rock salt sparingly over surface of floated concrete; embed, using float or piece of wood.

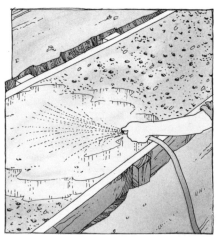

Hose out salt after covering and curing slab (see page 75).

after it has been floated, then press them in with a float or trowel and wait for the water sheen to disappear. Depending upon the smoothness you desire, finish by troweling or wood floating. After curing the slab, simply wash out the salt.

This finish is not recommended for areas with severe freezing weather. Water trapped in the pockets will expand upon freezing, and may crack or chip the surface.

Travertine finish. For a marbled effect, try the travertine finish.

After striking and floating, roughen the surface slightly with a broom—or just leave it very roughly floated—to prepare it for the next step. Using a large brush, dash a 1:2 cement-sand mix unevenly over the surface. Coloring the mixture (see below) to contrast with the concrete heightens the effect.

When the slab can support you on kneeboards, float or trowel the surface, knocking down the high spots. The result is a texture smooth on the high spots and rougher in the low spots. Cure the slab in the usual way.

Like salt finish, travertine finish is not resistant to severe freezing weather.

Simulated flagstones. One way to break up a dull expanse of plain concrete is to tool it so that it resembles flagstone.

Tool the concrete immediately after striking and floating. Use a concave jointer (see page 42) or bend an offset in a short length of ½ to ¾-inch copper pipe to make a good tool for this work. Sketch your pattern in advance and work from the plan. Erasures are awkward, so you need a sure hand.

When the water sheen has disappeared, do the final floating or troweling and redo the tooling. Trowel again if you want a very smooth finish. Touch up the surface with a soft brush and cure the slab.

Coloring concrete. Oxides can be applied by dusting them on a freshly placed concrete surface or mixing them with the concrete before placing. Greater economy results when you color only the surface layer. Colors are intensified when you substitute white cement for regular gray cement. Dusting is really for experts; amateurs tend to get splotchy results.

Masonry stains are simply painted on; results are less permanent than integral coloring.

TRAVERTINE FINISH

Dash mortar over surface of freshly floated slab, using a large brush.

Float or trowel mortar after it has stiffened somewhat; surface should have a stony look.

IMITATION FLAGSTONE

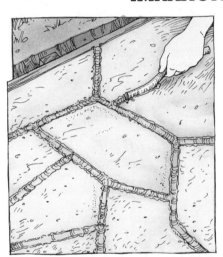

Tool the surface of a floated or trowelled slab, using a pipe or joint-finishing tool.

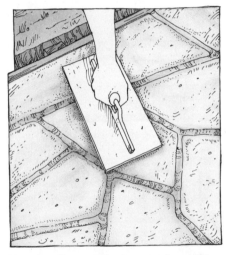

Trowel or float again, once surface stiffens; brush away crumbs and smooth out blemishes with trowel.

COLORING CONCRETE

Three ways to color concrete are shown above. At left, a coloring oxide is dusted onto a freshly floated slab. More floating or troweling will follow. Middle, a ½-inch layer of colored concrete is screeded over a fresh slab. At right concrete stain is brushed onto a cured slab.

Though masonry materials are fairly maintenance-free, they may require occasional cleaning, and in spite of their durability, they can be damaged.

The following section is your guide to the care and repair of masonry. Below, you'll find a discussion on the cleaning of masonry; a section on repairing brick, block, and poured-concrete masonry begins on the facing page.

CLEANING MASONRY

Most masonry can be kept clean with plain water, but you may occasionally have a problem that water can't cure. The following section will help you solve these problems.

Removing Efflorescence & Mortar Smears

Efflorescence is a white, powdery deposit caused by water dissolving the mineral salts contained in cement and brick; mortar smears are an inevitable result of learning to work with masonry.

Efflorescence. The mineral salts that appear as efflorescence, especially on brick paving, are carried to the surface by water, which then evaporates, leaving them behind. The deposits are harmless and will disappear once all the salts have been leached out. The catch is that this may take a couple of years. If you're impatient, read on.

Try brushing and scrubbing the deposits away, without using water; then follow with a thorough hosing.

Water tends to redissolve some of the salts, carrying them back under the surface, from which they will rise again. Remove as much as you can by dry scrubbing before turning on the hose.

For a stronger cure in extreme cases, follow the directions for removing mortar smears, below; it is a method that removes heavy deposits of efflorescence, too.

Mortar smears. Remove these with muriatic acid, available at masonry supply stores. The acid works by attacking the calcium contained in cement and mortar.

Use a 1:9 acid-water solution on concrete, concrete block, and dark brick. On light-colored brick, this solution may leave stains, so use a 1:14 or 1:19 solution. Do not use acid on colored concrete; it may leach out the color. Never use it on stone.

When preparing the solution, always pour the acid *slowly* into the water—*never* the reverse. Wear eye protection and rubber gloves, and work in a well-ventilated area. Apply the acid with a stiff brush to a small area at a time, let it stand for 3 or 4 minutes, then flush thoroughly with water.

Muriatic acid may change the color of masonry, at least slightly; bear this in mind when cleaning large surfaces. You may need to treat the whole area, even though mortar stains or efflorescence may affect only one part.

Removing Stains

Special remedies for stains on masonry are as varied as the stains themselves. For the most part, you can use ordinary household detergents, cleansers, and scouring powders. Specific remedies, and some cautions, are listed below.

Bear in mind that anything strong enough to really stain masonry has probably penetrated the surface. Complete removal of such stains is very difficult, if not impossible, so try some preventive maintenance.

Surfaces such as a brick kitchen floor or barbecue area should be sealed thoroughly with a commercial sealer. Sealers with a silicone base work best. You should apply 3 or 4 coats to ensure protection.

Do not use acids on stone, especially soft stone such as sandstone or limestone. These types should be cleaned with water only, as even detergents can be harmful. Always use fiber brushes; steel brushes are too abrasive and may leave rust marks.

Sometimes you can clean stone by rubbing it with a piece of the same rock, gradually abrading away the stain.

To clean concrete and brick, follow the directions below.

Oil & grease. Before the stain has penetrated, scatter fine sawdust, cement powder, or hydrated lime over the surface. If you catch it in time, these will soak up much of the oil or grease and then can be simply swept up.

If the stain has penetrated, try dissolving it with a commercial degreaser or emulsifier. These are available at masonry and home supply centers and at auto suppliers. Follow manufacterers' directions. Residual stains can sometimes be lightened with household bleach (see under "Rust" on facing page).

Avoid hazardous solvents such as kerosene, benzene, or gasoline; they aren't worth the risks of fire or toxic inhalation.

Paint. To clean up freshly spilled paint, wipe and scrub it up with a rag soaked in the solvent specified for the paint. For dried paint, use a commercial paint remover, following the manufacturer's instructions.

Rust. Ordinary household bleach will lighten rust stains (and most others). Scrub it in, let it stand, then rinse thoroughly.

A stronger remedy is a pound of oxalic acid mixed into a gallon of water. Follow the mixing directions for muriatic acid under "Mortar Smears," above. Brush it on, let it stand for 3 or 4 minutes, then hose it off. Remember that acid washes (and bleach) can affect the color of a surface. Test them in an inconspicuous area first.

Smoke & soot. Scrub with household scouring powder and a stiff brush, then rinse with water.

REPAIRING MASONRY

Cracked and broken bricks and blocks, crumbling mortar joints, and chipped and broken concrete are problems that may never occur in a well-made masonry structure. But shifting earth, impacts, and freeze-thaw cycles are beyond human control, and they can damage even a good mason's work.

This section guides you through the repair of brick, block, and stone structures (see below), tells you how to fix damaged poured concrete (page 82), and explains how to patch stucco (page 83).

Repairs of Brick & Block

Most trouble in a mortared wall or paving develops at the mortar joints. Sometimes the shrinking of modern cement-based mortar will cause the joints to open; old-fashioned lime-based mortar often just crumbles.

Freeze-thaw cycles worsen the problem. Water penetrates the tiniest cracks; upon freezing, it expands, enlarging the cracks and making it easier for the process to recur. Renewing the mortar joints will solve shrinking, crumbling, and cracking.

Settling of a mortared wall or paving will crack the joints, and sometimes the units themselves. Heavy impacts can do the same thing. This cracking calls for replacement of the mortar and possibly one or more units. In extreme cases, a whole section may need to be rebuilt.

Directions for all these repairs follow. The directions apply to all unit masonry walls and pavings—even though the drawings show only a brick wall and paving.

Renewing mortar joints. Fresh mortar will not adhere to old. Chisel out the cracked and crumbling old mortar with a hammer and narrow-blade cold chisel, exposing as much of the mortar-bearing faces of the units as possible. Expose the joints to a depth of at least ¾ inch, then thoroughly brush and blow them out. Always wear eye protection while using the chisel.

Dampen the area with a brush or a fine spray of water. Mix 1:2 cement-sand mortar (see pages 33–35) to a stiff consistency while waiting for the surface moisture to evaporate.

When the units are damp, but not shiny wet, use a joint filler or small pointing trowel to press mortar into the joints. You may find that a "hawk"—a small mortar board with a handle—will help you hold mortar close to the job.

Fill the joints completely, tamping the mortar in well (use a small piece of wood for deep joints). Masons call this "pointing." Tool the joints when the mortar is stiff enough (see page 42). Keep the repair damp for 4 days to cure the mortar.

Filling long cracks. You can fix long cracks by following the directions above for renewing mortar joints, but you'll probably find grouting easier.

Dampen cracked surfaces several hours before grouting. Then when surfaces are no longer shiny wet, grout can be poured into paving cracks from a bent coffee can, or into vertical cracks in walls through a funnel or a cardboard or tar

paper chute (see drawing below). Use 1:2 cement-sand grout (for more on mortar and grout, see pages 33–35).

Use wide, waterproof adhesive tape to dam up the grout in vertical cracks (see drawing below). If you have trouble getting the tape to adhere to the wall, hold it in place by bracing a board against it, as shown.

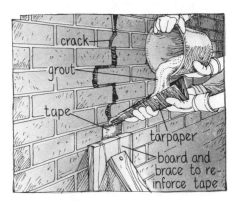

Fill 3 or 4 feet of crack at a time, waiting several hours between pours to let the grout set up. Keep the area damp for 4 days to cure the grout.

Replacing individual units. When a unit is badly damaged, you can replace it, provided it carries no load. Check by trying to move it; if it moves, you can take it out and replace it with a new brick or block. If it seems reluctant to move, it is probably carrying a load; leave it alone or consult a professional mason before removing it.

To replace a unit, chip out the old mortar with a narrow-blade cold chisel. Work carefully so as not to disturb adjacent units. Once the mortar is out, you should be able to remove the unit. If necessary, break it up carefully with the chisel and remove the pieces.

Clean up the cavity, removing all the old mortar. Wet the cavity and the replacement unit with a brush or fine spray of water, then mix a batch of 1:2 cement-sand mortar to the consistency of soft mud (see pages 33–35 for more on mortar).

When the cavity is damp but not wet, apply a thick layer of mortar to its bottom and sides and, if you're working on a wall, to the top of the new unit (see drawing below).

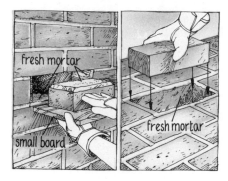

Push the new unit into place. A small board (shown) will aid in its alignment with a wall. Mortar should be squeezed from the joints; if not, add more. Trim off excess mortar and tool the joints (see page 42). Keep the area damp for 4 days to cure the mortar.

Rebuilding whole sections. Extensive damage calls for rebuilding. Whatever the extent of the damage, repair proceeds in the same way: work from the top down when taking damaged units out of a wall, and from the middle outward when fixing a paving. Reverse the process when you rebuild. Use a narrow-blade cold chisel to work on the mortar joints. Don't just bash away at the damaged units themselves; you may do further damage.

Replace the damaged units by following the instructions under "Replacing individual units," or, if the damage is extensive, by following the instructions for building unit masonry walls (pages 32–55), or for unit paving (pages 56–63). Tool the joints (see page 42) and keep the area damp for 4 days to cure the mortar.

Repairs of Poured Concrete

Concrete is hard and durable, yet if it is not placed, finished, and cured correctly, flaws can develop. Impacts, shifting earth, and freeze-thaw cycles can also take their toll.

This section guides you through a range of concrete repairs, from the correction of surface flaws and simple cracks through the repair of broken step edges to the rebuilding of heavily damaged slabs. Stucco is a cement-based material; a short section on repairing it concludes the chapter.

The security of any patch job on concrete largely depends upon the care you take in surface preparation. Always clean all dust and debris from the area to be repaired, and soak it thoroughly, even the previous day, before beginning work. There should be no standing water, but surfaces must be damp to ensure a good bond.

In any repair in which the patch will be thin or will need to be feathered at its edges, you'll find the extra strength of commercial latex or epoxy-cement patching compounds (sold under various trade names) well worth their extra cost.

Correcting surface flaws. If concrete is allowed to dry out too fast after placement or if it is finished excessively, surface flaws may develop. The most common flaws are *dusting,* in which the surface wears away easily; *scaling,* in which thin layers flake away from the surface; and *crazing,* in which fine networks of surface cracks appear.

Efflorescence, the appearance of white, powdery mineral salts on the surface, is not really a defect; it tends to occur naturally, and will disappear in time. To hurry it along, follow the directions under "Removing Efflorescence & Mortar Smears," page 80.

To correct dusting, scaling, and crazing, use one of the following methods: for light damage, clean the area and apply a 1:1 solution of linseed oil and mineral spirits. The oil will retard further damage and is also an aid in protecting concrete from deterioration due to the use of de-icing salts. You'll need to renew this coating every two years or so.

If the damage is bad enough to call for resurfacing, use either regular Portland cement mortar (1:3 cement-sand mix is good—see pages 33–35) or mortar to which a commercial epoxy or latex-based bonding compound has been added, especially for thin layers.

Prepare the surface by removing all loose and flaking concrete. A small sledge, used gently, will help break up scaling surfaces. Scrub the area clean and soak it as previously described.

Mix and apply the mortar according to the directions on pages 64–66, or if using epoxy or a latex-based mixture, follow the manufacturer's instructions. Float, trowel, or broom the surface of the patch to match existing surfaces and cure the patch carefully (see page 75).

Fixing cracks. Use a trowel to fill cracks up to about ⅛ inch wide with a stiff paste of cement and water or a commercial cement-based caulk.

Clean out larger cracks with a hammer and narrow-blade cold chisel, working to create a pocket ¾ inch deep or more for the repair material. Undercut the sides (see drawing below) to provide a positive "lock" for the patching material.

Prepare the area as previously described; then fill the crack with 1:3 cement-sand mortar mixed to the consistency of soft mud, or use a commercial latex or epoxy-cement patching compound, following the manufacturer's directions.

You can improve the bonding of mortar patches if you coat the area first with a thick "paint" of cement and water, scrubbing it in with a brush. Then apply the patch immediately—before the "paint" has a chance to dry. Commercial epoxy bonding compounds are used in the same way and form an even stronger bond.

Finish the repair to match adjacent areas, cover, and cure it carefully for 4 days. For more detail on concrete, see pages 74–75.

Repairing step edges. Here, as in "Correcting surface flaws" (page 82), you can choose either regular cement mortar or one of the commercial latex or epoxy-cement patching compounds.

If a piece has broken away from a step, simply cement it back into place using a masonry adhesive (follow the manufacturer's instructions). If the damage is more extensive, you'll need to chisel away the concrete until you have an undercut ledge of sound concrete that will support and retain a patch (see drawing below).

Premixed commercial latex or epoxy-cement patching compounds are often stiff enough to support themselves. However, if you use a regular cement mortar or are making large repairs, simple temporary formwork is a good idea. This can be nothing more than a board held against the step edge with blocks.

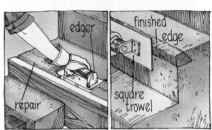

Thoroughly clean and then dampen the area several hours before placing the patch. Begin by brushing on a coat of thick cement-and-water "paint," or a commercial epoxy bonding compound. Follow it immediately with either a 1:3 cement-sand mortar (see pages 33–35) or a commercial epoxy-cement or latex patching compound. Thoroughly prod the material with a trowel to eliminate air pockets, then smooth it with the same trowel or a square steel finishing trowel (see drawing below).

When the patch has stiffened slightly, finish the edge with an edging tool (see drawing), then carefully remove the form board and use a trowel to touch up the face of the step. Cover and moist-cure the patch for several days (see page 75) or, if you are using a commercial compound, follow the manufacturer's instructions.

Mending heavily damaged slabs. Badly broken concrete slabs should be rebuilt with more concrete. If the damage is extensive, this may mean erecting forms and pouring a new slab (see pages 74–75).

Begin by breaking up the damaged area with a sledge; save the pieces to use as filler in the repair.

If the gravel base has sunk, build it up with more gravel and bits of broken concrete. Clean the edges of the patch area and saturate everything thoroughly with water. Wait several hours for the standing water to be absorbed before placing the fresh concrete. If the broken area extends to the slab edge, erect temporary forms (see drawing below).

Mix a batch of 1:3:3 cement-sand-aggregate concrete sufficient to fill the area, and place, finish, and cure it according to the directions found on pages 74–75.

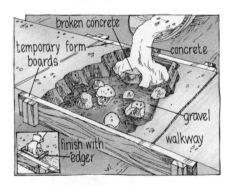

If the slab must carry heavy loads—a driveway, for example—you'll need to add reinforcing, locking the repaired area to the old slab. To do this, drill holes in the exposed edges before adding the new concrete (see page 93 for more on masonry drills) and cement pieces of reinforcing rod or large bolts in with mortar (see pages 33–35) or commercial masonry adhesive. Heavy wire mesh reinforcing can then be tied to the rods or bolts with wire; support the mesh in the center with bits of the broken-out concrete. Then rebuild as described above.

Patching Stucco

You can repair damage to stucco walls with 1:3 cement-sand mortar to which a maximum of 1/10 part hydrated lime has been added (see pages 33–35). Or you can use a commercial stucco-patching compound; for small jobs, this is the most convenient and economical way.

Begin by removing all damaged material with a cold chisel or a putty knife until you get down to a bed of sound stucco (see drawing below). You may expose the lath or stucco netting underlying the stucco; in this case, plan to make the repair in two layers.

Scrub the area clean and moisten with a brush or fine spray of water. Mix the mortar or patching compound to a workable but not runny consistency. Apply the material with a mason's trowel or putty knife, smoothing or texturing it to match the original surface. If you are making a two-layer repair, cure the first layer for at least 2 days before adding the second.

Matching color and texture is ticklish, but here are some hints. If the wall is to be painted, use regular gray cement in the mortar. If the color is integral, see your supplier for coloring oxides, and plan to use white cement and sand in the mortar mix. Textured surfaces can be produced with a wood float (see page 75), a sponge, or a piece of carpet.

To cure the repair, keep it moist for 4 days by misting it with water. If the weather is windy or hot, or if the area is in strong sunlight, cover it with plastic sheeting or damp burlap to help retain the moisture.

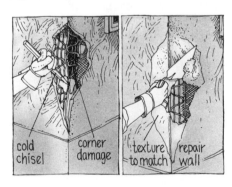

MASONRY PROJECTS
In & Around the Home

In the following chaper, you'll find a collection of project
ideas that show you how to use masonry to dress up your home. Each project is
general enough in nature to be adaptable to your own
situation and is accompanied by detailed drawings. The drawings feature prominent
labels that refer you back to basic information in the earlier
how-to chapters—information you'll need to complete the specific project illustrated.

STEPS & STAIRS

On these two pages you'll find ideas that demonstrate how masonry can ease your ups and downs around the garden. Some experience in working with concrete and brick is advisable.

Materials. Your choice of materials should depend more on your budget and the effect you want to create than on durability—all masonry materials make long-lasting steps.

Consider stone for its rustic appeal or brick for its range of effects, from rustic to formal. Poured concrete is excellent for utility and can be given an attractive exposed-aggregate finish. Concrete blocks make simple, inexpensive steps and can serve as a foundation for other materials. Adobe and tile lend their own special character.

Proportions. Architects and builders long ago worked out a set of ideal proportions for steps: twice the riser height (each step's vertical dimension), plus the tread depth (each step's horizontal dimension) should equal 25 to 27 inches. A comfortable average is a 6-inch riser with a 15-inch tread. Risers should be at least 4 inches high, but no higher than 7 inches. Treads should be no shorter than 11 inches (see drawing above right).

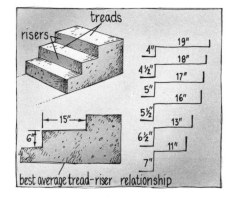

best average tread-riser relationship

Dimensions. Calculate the change in level of your slope using the simple device shown in the drawing below. The distance from A to B is the change in level (also known as the "rise"); the distance from A to C is the "run"—the minimum distance your steps will run.

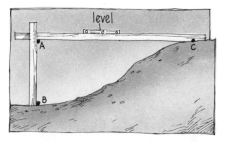

Determine the number of steps you'll need by dividing the desired riser height into the total rise of the slope. Check the table under "Pro-

portions," above, to see if the corresponding minimum tread will fit into the slope's total run. You will probably have to excavate—cut the bank—or fill under the step to make the slope fit the stairway.

Plan on a minimum width of 2 feet in a simple utility stair. Four feet is a good minimum width for a more gracious stair; make that 5 feet if you want to enable two people to walk abreast.

Planning. Take the time to make a scale drawing—you don't have to live with a bad drawing, but a masonry project gone awry is a different matter. Plans for steps that abut a sidewalk or other public access may have to be cleared with your building department—another reason to make a drawing.

Construction. Steps with bulkheads formed by railroad ties or other timber are the easiest to construct (see facing page, top). Poured-concrete steps involve the use of a stepped form (see page 73 and drawing on facing page). The same type of form can be used to pour a concrete base for unit paving—the lower drawing on the facing page shows one example, using bricks.

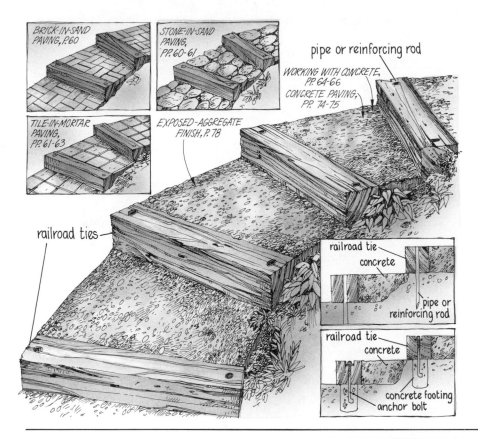

BRICK-IN-SAND PAVING, P. 60
STONE-IN-SAND PAVING, PP. 60-61
TILE-IN-MORTAR PAVING, PP. 61-63
EXPOSED-AGGREGATE FINISH, P. 78

pipe or reinforcing rod

WORKING WITH CONCRETE, PP. 64-66
CONCRETE PAVING, PP. 74-75

railroad ties

railroad tie
concrete

pipe or reinforcing rod

railroad tie
concrete

concrete footing
anchor bolt

RAILROAD-TIE STEPS

Railroad ties make for simple, rugged steps. To build, excavate for your steps, compacting the soil in the tread area firmly. Then simply pin the ties in place with pieces of reinforcing bar or pipe set in drilled holes. You can drive the steel into the earth with a sledge. For extra security, provide small poured-concrete footings as shown. Holes for the concrete also serve as forms—they are made with a post-hole digger. Set anchor bolts in the slightly stiffened concrete. When the concrete has set (after about 2 days), you can bolt the ties to the footings.

Once the ties are in place, fill the tread spaces behind them with poured concrete (page 74), brick-in-sand paving (page 60), or any other material that suits your fancy.

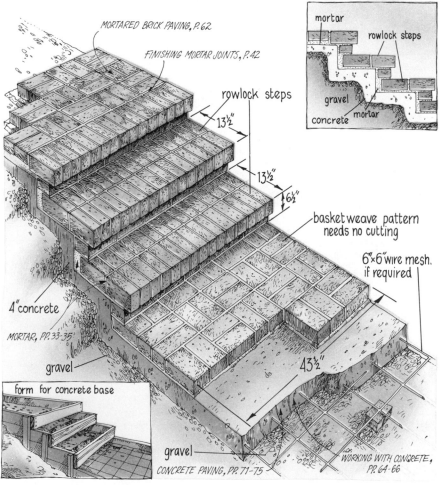

MORTARED BRICK PAVING, P. 62
FINISHING MORTAR JOINTS, P. 42

mortar

rowlock steps

gravel

concrete   mortar

rowlock steps

13½"

13½"

6½"

basketweave pattern
needs no cutting

6"×6" wire mesh,
if required

4" concrete

MORTAR, PP. 33-35

gravel

43½"

form for concrete base

gravel

CONCRETE PAVING, PP. 71-75

WORKING WITH CONCRETE, PP. 64-66

BRICK-ON-CONCRETE STEPS

The drawing at left shows a typical brick stair. You can change proportions just by referring to the chart on the facing page. Construction involves moderate skill in pouring concrete and in bricklaying; read the chapters beginning on pages 32 and 64 before assembling your materials.

Begin by excavating and tamping the soil base for the concrete. Spread a 4-inch-thick gravel layer to promote drainage and guard against frost damage. Assemble simple 1 by 4 plank forms as shown in the detail drawing (see pages 71–73 for complete information on form building), then pour, finish, and cure the concrete according to the directions on pages 74–75.

Lay the bricks in mortar according to the unit paving instructions found on page 62. The drawing shows brick-in-mortar paths in a simple basketweave bond pattern at both ends of the steps. For more bond patterns, see page 56.

VENEERS

Veneers are among the most popular applications of masonry. A veneer saves expensive masonry materials by using them only on a project's visible surface, hiding a core of less attractive — and less costly — material.

On these two pages you'll find three typical veneer applications: brick on a frame house, stone on concrete block, and synthetic stone on an interior fireplace wall. These applications are adaptable to many home projects.

BRICK ON A FRAME HOUSE

Brick applied to exterior house walls is a popular way of dressing up a house. It requires quite a bit of masonry expertise. The veneer may cover the wall from roof to foundation, as in the left half of the drawing, or it may stop at window sill level, as in the right half.

The drawing at right is intended only as a guide to the building of a typical veneer wall. You must obtain a building permit before beginning work. Your building department may require variations.

From the bottom up, here's how the veneer is added. Support is provided by a corrosion-resistant steel angle bolted through the existing foundation, or by a small footing poured next to the house foundation. The existing siding is then covered with a layer of tarpaper.

The wall is laid up in a single wythe, or tier. Running bond or stack bond is usually used. The veneer is held in place with metal wall ties nailed through the siding into the studs. The ties should be spaced every 32 inches horizontally and every 16 inches vertically, with the rows offset so that the ties do not line up. When laying the bricks, adjust and maintain the mortar joint thickness precisely to secure the desired wall height. A story pole (see page 34) helps you maintain accuracy.

The veneer is flashed to help prevent water from seeping behind it, and it is provided with a drainage system so that any water that seeps in in spite of the flashing can escape. Do not neglect these details; trapped water can seriously damage the structure of the house.

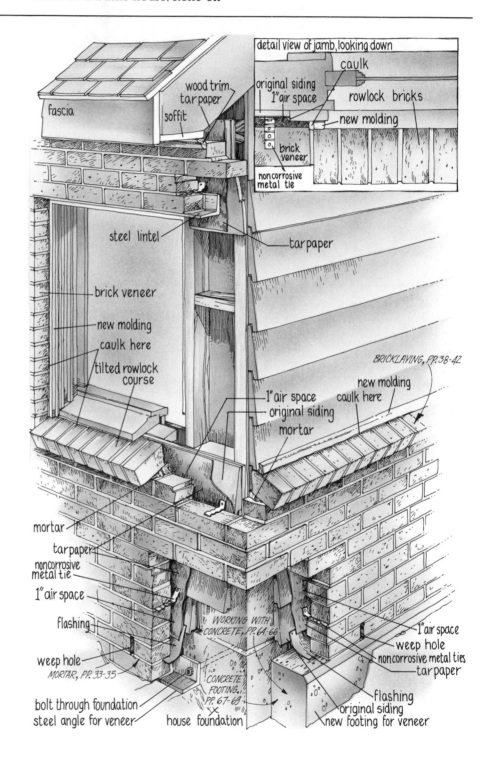

detail view of jamb, looking down
caulk
original siding
1" air space
rowlock bricks
new molding
brick veneer
noncorrosive metal tie

wood trim
tar paper
fascia
soffit

steel lintel

tar paper

brick veneer

new molding

caulk here

tilted rowlock course

1" air space
original siding
mortar

BRICKLAYING, PP. 38-42

new molding
caulk here

mortar
tarpaper
noncorrosive metal tie
1" air space
flashing
weep hole
MORTAR, PP. 33-35

WORKING WITH CONCRETE, PP. 64-66

bolt through foundation
steel angle for veneer
house foundation

CONCRETE FOOTING, PP. 67-69

1" air space
weep hole
noncorrosive metal ties
tar paper
flashing
original siding
new footing for veneer

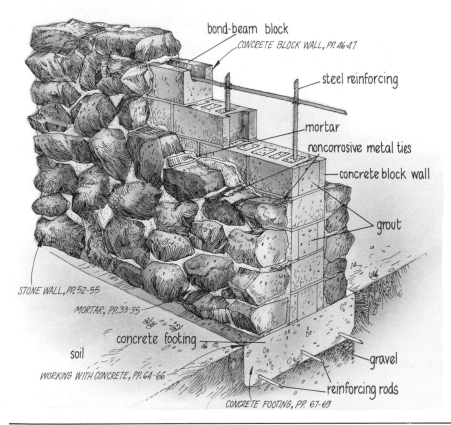

bond-beam block
CONCRETE BLOCK WALL, PP. 46-47

steel reinforcing

mortar

noncorrosive metal ties

concrete block wall

grout

STONE WALL, PP. 52-55

MORTAR, PP. 33-35

concrete footing

soil

WORKING WITH CONCRETE, PP. 64-66

gravel

reinforcing rods

CONCRETE FOOTING, PP. 67-69

STONE ON CONCRETE BLOCK

One of the best ways to dress up a plain concrete block wall is to veneer it with stone. The result appears to be a stone wall, but can be achieved at much less labor and expense.

Use masonry nails, screws in fiber plugs, or a stud gun to attach wall ties to the block wall every 2 or 3 square feet (see "Masonry Fasteners" page 93). If you are building the concrete block core wall from scratch, insert the ties in the mortar joints between blocks.

The veneer stones are attached to each other and to the concrete block wall with mortar, the wall ties providing a positive connection to the mortar. Bend as many of the ties as possible into the joints between stones. "Slush fill" the space between wall and stones completely with soupy — but not runny — mortar as you go.

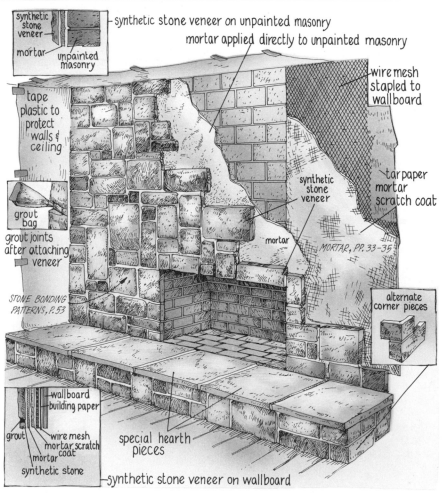

synthetic stone veneer
mortar
unpainted masonry

synthetic stone veneer on unpainted masonry
mortar applied directly to unpainted masonry

wire mesh stapled to wallboard

tape plastic to protect walls & ceiling

synthetic stone veneer

tar paper mortar scratch coat

mortar

MORTAR, PP. 33-35

grout bag

grout joints after attaching veneer

alternate corner pieces

STONE BONDING PATTERNS, P. 53

wallboard
building paper

grout

wire mesh
mortar scratch coat
mortar
synthetic stone

special hearth pieces

synthetic stone veneer on wallboard

SYNTHETIC STONE ON INTERIOR WALLS

Lightweight synthetic stone veneer has become increasingly popular as a means of dressing up interior walls. The veneer can also be used on exterior walls. Made of a cement-based mixture, this veneer can be startlingly realistic, and is easy to work with.

The drawing at left shows a typical veneer installation. Always follow manufacturer's instructions; the procedure shown in the drawing is only meant to give you an idea of what's involved.

The "stone" can be mortared directly to unpainted masonry walls; walls made of other materials must receive first a layer of building paper and then one of chicken wire or metal lath. An initial coat of mortar is applied over this and scratched to provide a good "lock" with the top coat of mortar. After the scratch coat has cured, the "stone" is mortared in place over it. Joints are either kept tight or filled by means of a grout bag (see drawing).

RETAINING WALLS

Many municipalities require a permit for any retaining wall, since the quality of construction and design is critical, and many also specify that walls over 4 feet high must be designed and supervised by a licensed engineer. Thus you *must* consult your building department any time you contemplate a project on a more ambitious scale than a low garden terrace, planting bed, or tree well.

The discussion that follows is designed to introduce you to the subject of retaining walls. On the facing page are two examples of retaining walls built to typical specifications. These walls may be fine for your community; on the other hand, they may not meet your local specifications.

Preparing the Slope

Even the gentlest of slopes requires some alteration before a wall can be built, and appropriate planting helps control erosion.

Cutting & filling. The drawing below shows three typical methods of site preparation. In the first, the slope is cut away and the earth moved downhill to create a plateau. In the second, a raised bed is created by cutting away below the wall site and filling on the uphill side. The third method divides the total wall height into two smaller walls, creating terraces—usually the best idea wherever possible.

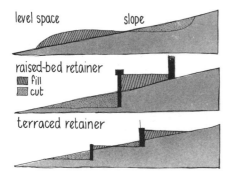

Planting. Hardy, firm-rooted plants that cover well but won't spread too fast to control will help to retain the soil. Plant right away after completing the wall, while the soil is still soft and easily worked.

Water Control

Drainage is an essential part of retaining-wall design. Once rains have saturated the soil, water flows downhill, both above and below the ground. Your retaining wall acts like a dam, restricting this flow of water. You must make some provision for drainage or risk undermining or even bursting the wall.

If you are dealing with expansive clay soil, you have an additional problem. The saturated soil expands and acts like a hydraulic jack, pressing against the wall. In this case, professional engineering is called for.

Drainage. The drawings on the facing page show two typical drainage schemes. At top, water is collected in gravel backfill and led around the edges of the wall. If you opt for this design, be sure to funnel the resulting concentration of water toward a storm sewer, ditch, or natural drainage.

The bottom wall incorporates weep holes that pass through the wall.

Both drawings show tarpaper covering the gravel backfill. This prevents the soil from sifting down from above into the gravel and clogging it.

Retaining-wall Designs

There are two basic types of retaining-wall designs: mass and cantilever. A mass wall, such as the one in the drawing at the top of the facing page, is held in place by simple pressure of its mass; gravity does the work.

A cantilever wall relies on the strength of steel reinforcing. The cantilever design shown on the facing page, bottom, has a very wide foundation to help it resist toppling and "sledding"—the tendency of earth pressure to cause it to slide outward.

Retaining walls can be built from a variety of materials, though engineering must take precedence over appearance in most cases. Here are the pros and cons of the masonry materials most often used in retaining walls.

Poured concrete. Strongest of all, concrete walls stand where others fail. The cost of formwork can become prohibitive for residential projects larger than the poured wall shown on the facing page, however.

Concrete block. Solid-grouted, reinforced concrete-block walls are about as strong as the poured-concrete kind; the units act as permanent forms for the poured concrete. This is usually the most economical form of construction, small or large. Appearance can be enhanced by veneering (see page 87) or by the use of sculpture-face blocks and bond-pattern variety (see pages 44–45).

Brick. Brick retaining walls are usually laid up in two tiers, or wythes, with steel reinforcing between them. Grout fills the cavity, locking in the steel, and strengthening the wall. Often the inner wythe is built of inexpensive concrete blocks, which make for speed of construction as well as cost savings. Brick walls are weaker than the concrete varieties and are best considered for appearance.

Stone. Thick, mortared stone retaining walls are stronger than brick ones, mostly because of the extra thickness. Dry-laid stone walls have a rough-hewn beauty and make attractive retainers. As the drawing below shows, dry-laid walls should be tilted back into the slope— "battered," as masons would say—to increase their holding power. The joints can be planted to enhance the wall's appearance.

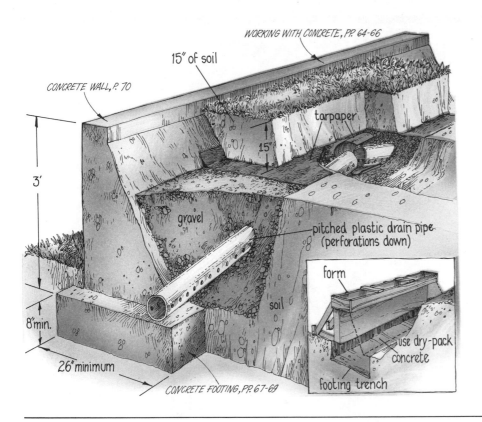

WORKING WITH CONCRETE, PP. 64-66

15" of soil

CONCRETE WALL, P. 70

tarpaper

15"

3'

gravel

pitched plastic drain pipe
(perforations down)

form

soil

use dry-pack
concrete

8" min.

footing trench

26" minimum

CONCRETE FOOTING, PP. 67-69

POURED-CONCRETE MASS WALL

The wedge shape of this mass wall helps hold it in place, and also simplifies formwork, as the detail drawing shows. The width of the base must be ½ to ¾ the wall height. At the indicated height, steel reinforcing usually is not needed, but a single ⅜-inch reinforcing bar can be placed horizontally near the top for extra security.

Dry-pack concrete simplifies construction: a form is built for the vertical face, then fairly stiff concrete is packed in, liberally laced with stones or broken concrete. (Concrete should be stiff enough to hold its shape in the shovel.) Whacking with a shovel compacts the mix; if the mix is stiff enough, it will hold its shape on the sloping side without slumping, and the entire wall can be completed in a single pour.

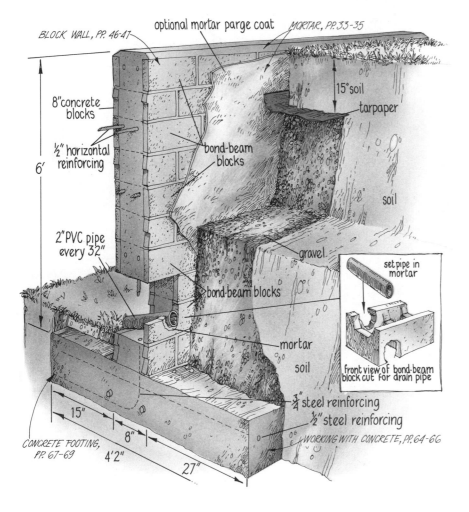

BLOCK WALL, PP. 46-47

optional mortar parge coat

MORTAR, PP. 33-35

15" soil

8" concrete
blocks

tarpaper

½" horizontal
reinforcing

bond-beam
blocks

6'

soil

2" PVC pipe
every 32"

gravel

bond-beam blocks

set pipe in
mortar

mortar

soil

front view of bond-beam
block cut for drain pipe

⅜" steel reinforcing
½" steel reinforcing

15"

8"

*CONCRETE FOOTING,
PP. 67-69*

4'2"

27"

WORKING WITH CONCRETE, PP. 64-66

CONCRETE BLOCK CANTILEVER WALL

This high retaining wall is built of concrete blocks for economy and ease of construction. Its steel reinforcing meets most codes but may not meet requirements in all areas, so be sure to check with local building officials.

The foundation is poured first, with steel reinforcing incorporated. (For more on reinforcing, see page 68.) Then the wall is erected, using bond-beam blocks and horizontal reinforcing in every other course. Every other block in the lowest bond-beam course is notched to receive plastic pipe, as shown in the detail drawing. The pipes form weep holes for drainage. Grouting is done in stages as each bond-beam course is completed. A mortar cap completes the project.

A ¾-inch coat of mortar, called a "parge" coat, trowelled onto the back of the wall, will help to control dampness on the face. Heavy tarpaper will do the same thing.

Retaining Walls

POOLS & PLANTERS

Masonry can be used to make fine containers for water or plants. On these pages are designs for both — a small garden pool and two planters, one made of brick, the other of poured concrete. As with most masonry projects, you can adapt these to suit your space and needs. All three designs are good for beginners and can be made in almost any size and shape, so don't feel limited by the scale shown here. Think of these projects as points of departure.

GARDEN POOL

First, excavate and compact the area. Allow for a finished depth of at least 18 inches if you plan to stock the pool with fish; this will keep them safe from cats, racoons, or other marauders. In areas of severe freezing weather, allow an additional 3 to 4 inches for a gravel underlayer. Slope the sides at 45 degrees. Allow for the shallow perimeter lip that will support the stone coping.

Add reinforcing either by bending 6 by 6 wire mesh to fit inside the pool, or by using ¼-inch or ⅜-inch reinforcing rods. Bend rods to follow the pool contours, and arrange them so that they resemble latitudinal and longitudinal lines. Space them 6 to 12 inches apart, tying the intersections securely with wire.

Using bits of brick or stone, support the reinforcing 2 inches off the earth or gravel. Then drive stakes in every square foot.

Mark the stakes 4 inches above the earth or gravel. Mix 1:2:3 cement-sand-gravel concrete with just enough water to wet all ingredients. Using a shovel or trowel, pack the mix firmly around the reinforcing up to the marks on the stakes. Remove the stakes and fill the holes with concrete. Finish the surface with a float or square trowel, then cover and cure the shell (see page 75).

Finally, add a stone coping, mortaring the stones to each other and to the concrete lip. A small waterfall, as shown, adds a nice touch. A small, submersible, recirculating pump moves the water and provides a means of draining the pool. Route its power cord and supply tube up behind the waterfall. For more on ornamental pools, see the *Sunset* book, *Garden Pools*.

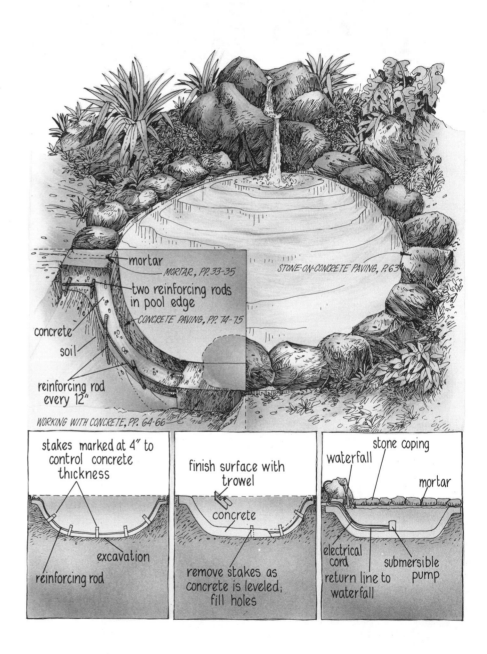

mortar

MORTAR, PP. 33-35

STONE-ON-CONCRETE PAVING, P. 63

two reinforcing rods in pool edge

CONCRETE PAVING, PP. 74-75

concrete

soil

reinforcing rod every 12"

WORKING WITH CONCRETE, PP. 64-66

stakes marked at 4" to control concrete thickness

excavation

reinforcing rod

finish surface with trowel

concrete

remove stakes as concrete is leveled; fill holes

stone coping

waterfall

mortar

electrical cord

return line to waterfall

submersible pump

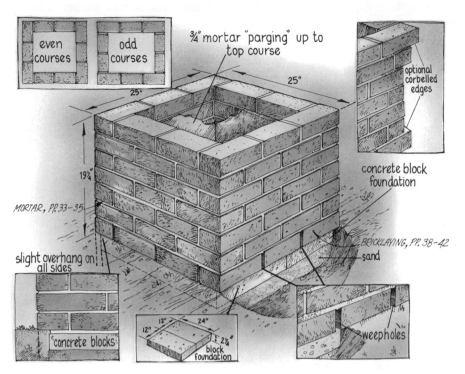

even courses

odd courses

¾" mortar "parging" up to top course

25"

25"

optional corbelled edges

concrete block foundation

19¼"

MORTAR, PP. 33-35

BRICKLAYING, PP. 38-42

sand

slight overhang on all sides

"concrete blocks"

12"

24"

12"

24"

2¼"

block foundation

weepholes

SIMPLE BRICK PLANTER

This easy-to-build brick planter sits atop two 12 by 24 by 2¼-inch concrete blocks—used as cap blocks for walls and as pavers, widely available at building centers.

Set the blocks below grade, levelled on a 2-inch-thick bed of sand. Then lay out a dry course of bricks, marking mortar joints with a pencil. Lay up the bricks and tool the mortar joints according to the directions on pages 38–42. Make weep holes for drainage by leaving out vertical mortar joints in the first course.

Apply a "parge" coat of mortar to the inside of the planter to aid in waterproofing (see drawing). You can allow the top courses to overhang slightly (see detail, upper right) to add visual interest. This is known as "corbelling."

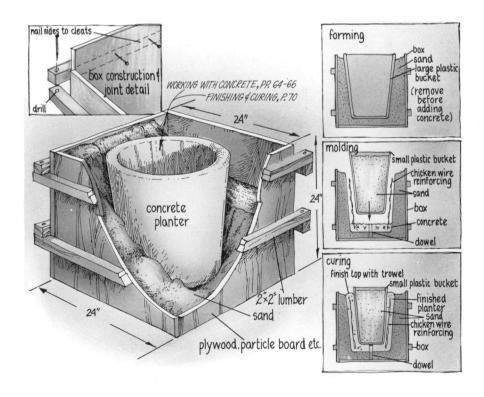

nail sides to cleats

box construction & joint detail

drill

WORKING WITH CONCRETE, PP. 64-66
FINISHING & CURING, P. 70

24"

24"

concrete planter

24"

2×2" lumber

sand

plywood, particle board etc.

forming

box
sand
large plastic bucket
(remove before adding concrete)

molding

small plastic bucket
chicken wire reinforcing
sand
box
concrete
dowel

curing

finish top with trowel
small plastic bucket
finished planter
sand
chicken wire reinforcing
box
dowel

SAND-CAST CONCRETE PLANTER

This small planter is cast in a sand mold formed inside a knockdown box. The plans can be adjusted to almost any size you like.

The outer surface of the planter is formed by packing damp sand around a plastic bucket, then removing the bucket (step 1, right). Or you can form an irregular hole in the sand with your hands, to yield a free-form planter.

In step 2, 2 inches of concrete are placed in the bottom of the hole, followed by a greased dowel, which forms a drain when removed after the concrete has cured. A reinforcing cylinder of chicken wire follows immediately, then a smaller plastic bucket, to form the inner surface of the planter.

In step 3, concrete (1:3:3 mixture of cement, sand, and pea gravel) is poured in after placing the chicken wire, to form the planter; it is then covered and cured (see page 75). The small bucket can be removed after a day, and the whole assembly can be knocked down two days later.

BARBECUES

Masonry has been the traditional material for built-in-place barbecues. Prefabricated grills, doors, and cook units have made the do-it-yourselfer's job easier; now, all you have to do is provide a masonry shell.

Here are two designs that take advantage of prefabricated parts. Always buy these parts first, then use them as the basis for the barbecue's dimensions. For more information, and many more plans, see the *Sunset* book *Ideas for Building Barbecues*.

RUSTIC GRILL

The drawing at right shows a barbecue built of cobblestones, though it could as easily be made of any unit masonry—even poured concrete. It's a good project if you're trying your hand at masonry for the first time.

Dimensions of this barbecue are based upon its 18 by 24-inch grill. Firebricks line the floor and sides of the barbecue and also support the grill. They are assembled with 3:6:1 cement-sand-fireclay mortar. The whole project sits on a simple concrete footing. The stonework is mortared to the footing, but the firebrick floor is held in place by the surrounding earth.

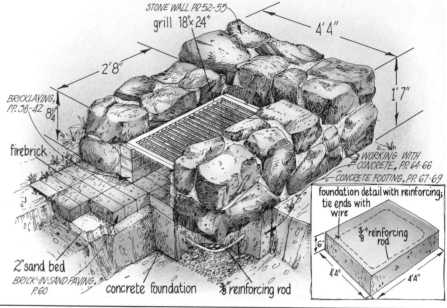

GRILL & COUNTER

The brick barbecue shown at right is a project for experienced bricklayers. The design can be adapted for small spaces just by lopping off one or both of the storage units. The drop-in cook unit, the tool rack, and the cutting board can be removed and stored away for the winter.

The building of this barbecue must be precise. The prefabricated doors are set in place as the work proceeds, and their frames are held in by the pressure of the surrounding bricks, along with a little mortar.

The stone top should be cut for you at a stone yard; it's simply mortared in place. If stone is unavailable you can cast a 3-inch-thick poured-concrete slab top in place on the barbecue (see drawing, top right). Use plywood on temporary supports to form the bottom of the slab and a collar of 1 by 4 lumber to form its edges. Two layers of wire mesh reinforcing, spaced evenly in the slab's thickness, should be used.

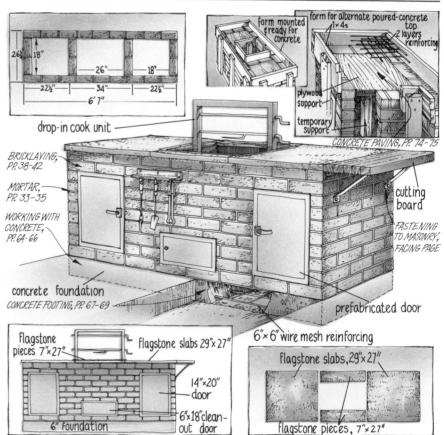

MASONRY FASTENERS

Sooner or later, you'll probably need to fasten something to masonry.

Expansive Anchors

Within the broad category of expansive anchors, you'll find everything from simple fiber plugs to fancy expansion nuts. They all work on the same principle: a sleeve of fiber, plastic, or metal is inserted in a hole drilled in the masonry. Threading a screw or bolt into the anchor (or sometimes just driving in the anchor itself) causes the sleeve to expand against the walls of the hole, holding it fast.

To use any expansive anchor, you'll need to drill a hole first. Star drills come in a variety of sizes; the smaller ones mount in a special steel holder. You use this type of drill by striking it with a hammer, turning the drill a few degrees between blows. Some star drills are designed for use with a special impact tool which facilitates the drilling process. The finished hole tends to be a little larger than the drill size.

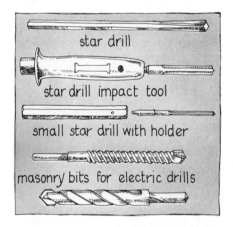

star drill

star drill impact tool

small star drill with holder

masonry bits for electric drills

Masonry drill bits differ from star drills in that they are usually carbide-tipped and mounted in an electric drill. Masonry bits can also be used with impact drills—tools that greatly speed the drilling operation by combining rotary motion with rapid hammering.

Once the hole is drilled, you're ready to install the fastener. Here is a list of the ones you'll find when you visit your masonry supplier.

Fiber plugs. Simple and inexpensive, fiber plugs are meant for light loads and small screws (see drawing below). Pick a plug that matches the screw you are using, and drill a hole as specified by the plug manufacturer. Then insert the plug in the hole, pass the screw through the fixture you are attaching, and thread it into the plug. The plug will expand against the sides of the hole and will remain in place should you later remove the screw.

Lead and plastic anchors. These work just like fiber plugs, but are meant for somewhat heavier loads (see drawing below). Use a plastic anchor when a fiber plug isn't strong enough, and a lead anchor when a plastic one is insufficient; your supplier can help you decide. Install the anchor just as you would a fiber plug (see above).

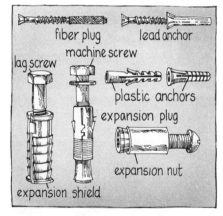

fiber plug   lead anchor

machine screw

lag screw

plastic anchors

expansion plug

expansion nut

expansion shield

Expansion shields, plugs, and nuts. For heavy-duty fastening, these devices are the answer (see drawing above). Expansion shields have a split body that, inserted into the drilled hole, expands against the hole when a lag screw is threaded in. Expansion plugs contain a conical wedge. When a machine screw is threaded in, it draws the conical wedge up into the body of the plug, expanding the body. A variation on the same idea is the expansion nut. To install one, you drill the hole to a precise depth so that the conical wedge at the bottom of the nut will contact the bottom of the hole. The nut is installed by hammering on a special tool that drives the nut against the bottom of the hole, forcing the wedge up into the body so that it locks the nut in place. Expansion plugs and nuts take machine screws and provide more permanent threads in the masonry than lighter plugs.

Masonry Nails

Special tempered-steel nails are available for such typical jobs as the attaching of furring strips to a masonry wall. Always wear safety glasses when working with masonry nails. If you hit a nail wrong, it won't bend; it will break, sometimes scattering bits of shrapnel around the room with considerable force.

A driving tool is safer. It has a handle with a sliding steel striker in one end and a hole in the other (see drawing below). To use it, you place a special drive pin in the hole, set the tool against the workpiece, and drive the pin through the workpiece into the masonry by striking the steel striker with a small sledge or heavy hammer.

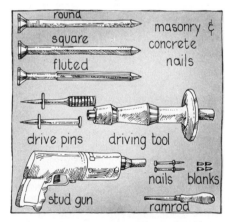

round

square

fluted

masonry & concrete nails

drive pins   driving tool

nails   blanks

stud gun

ramrod

Stud guns can be rented. A stud gun literally fires a nail (by means of a .22-caliber blank cartridge) through the workpiece and into the masonry (see drawing above). You adjust the force of the impact by selecting among a range of powder charges—light to heavy. A safety prevents the gun from firing except when pressed firmly against the workpiece. Thoroughly discuss your needs with your supplier, and always follow the manufacturer's instructions to the letter when using the gun, to avoid accidents.

REMODELING WITH MASONRY

The house at right combines the bare-bones features typical of many houses, from tract homes to custom-built residences: a driveway that provides the only approach to the house; an entry path too narrow for two to walk abreast; a featureless concrete porch; a patio slab in the rear too small for entertaining; and an unpaved sideyard.

As a glance at the drawing on the facing page will show, it's possible to work quite a transformation on this house and garden through the use of masonry.

The Sideyard

An unpaved sideyard is a nuisance, but one that is easily solved with a straightforward, functional concrete paving. The one shown in the drawing on the facing page is 3 feet wide—wide enough for utility, but narrow enough for ease of construction. Control joints (see page 74) are formed every 4 feet; a simple, wood-float finish provides traction.

Begin by grading the area to be paved (see page 58), then set temporary edgings (see pages 71–72). Place, finish, and cure the paving according to the directions on pages 74–75.

The Back Yard

A good place to begin is with concrete steppingstones—really small concrete slabs—that connect the sideyard and patio. These are cast in place using a multiple-grid form (see page 77). The width is 3 feet, with lengths varying to make a pleasing pattern. An exposed-aggregate finish (see page 78) is both beautiful and safe underfoot.

Extending the patio. Small slab patios like the one shown can be a headache. Often they are too small to be of much use, but there is a graceful way to extend them.

The answer lies in the brick paver—a brick only half as thick as a regular brick. Split pavers are even thinner. Choose a kind that matches a full-thickness brick, and

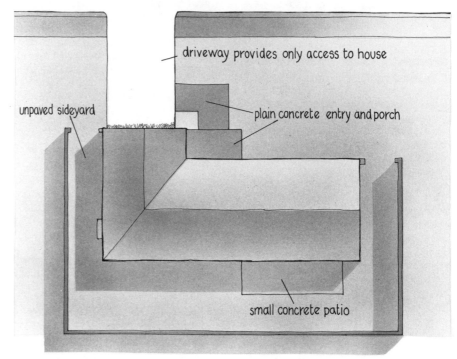

driveway provides only access to house

plain concrete entry and porch

unpaved sideyard

small concrete patio

you can veneer the concrete with the pavers, merging them seamlessly into a new full-thickness, brick-in-sand paving.

The pavers are fastened to the concrete with a thin coat of tile-setting adhesive, available at masonry and tile suppliers. The new brick-in-sand paving is retained by "invisible" edgings (see page 59). The choice of a basketweave pattern, with its straight edges, simplifies the potentially awkward joint between the two pavings (see drawing below).

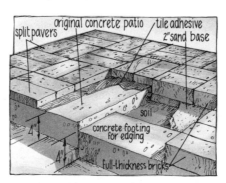

split pavers   original concrete patio   tile adhesive   2" sand base

soil

4"   concrete footing for edging

4"   full-thickness bricks

Beginning the veneer at the patio edges, work back toward the house with the pavers, following the directions supplied with the adhesive. You'll probably need to cut the pav-

ers to fit them along the house wall; see page 39 for instructions.

Begin the brick-in-sand paving by making the edgings (see page 59). Then place the sand bed and set the paving according to the directions beginning on page 60. It's a good idea to set the paving ¼ to ½ inch too high to allow for settling. Joints between both bricks and pavers should be butted and sanded.

Wall, barbecue & pool. Add a low seating wall by following the directions for brick walls on pages 38–42. Walls of other materials are possible, too. Whatever material you use, make the wall at least 12 inches thick. If brick is used, this complicates the bonding pattern slightly (see page 43). A minimum height of 16 inches is best for seating.

Consider adding a barbecue or an ornamental pool. Instructions for these projects begin on pages 92 and 90. See also the *Sunset* books *Ideas for Building Barbecues* and *Garden Pools*.

The Entry

The same combination of brick paver veneer and brick-in-sand paving can be used to dress up a plain entry. The design widens the path to your door, and gives you room to step out of your car without setting foot on a damp lawn.

Follow the instructions for extending the patio, above, but with one major difference at the entry path. If you veneer the path just as you would the porch, it will result in a step just high enough at the driveway edge to trip you up.

A better solution is shown in the drawing below. Install the brick pavers on the porch (and any steps) first. Then build low walls on the path to contain a bed of concrete and a full-thickness brick paving with sanded joints. Make the height of the walls equal to the height of the porch plus its new pavers (or the height of the lowest step plus its new pavers).

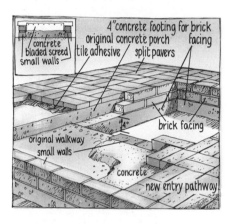

Build the walls according to the bricklaying instructions on pages 38–42, using the old path as the footing. Then screed a layer of concrete sufficiently deep to bring the finished brick surface up to the height of the walls. Lay the bricks on the cured concrete, and sand the joints as you would for brick-in-sand paving (page 60).

Lay out and install the rest of the new entry path according to the instructions for brick-in-sand paving (pages 58–60). The one shown uses an "invisible" edging (see page 59), the foundation of which is poured in curved forms made of redwood benderboard.

When pouring the footing for the edging, add a small footing along the edge of the porch to support a mortared brick facing that will hide the porch's last bit of concrete, as well as the paver edges.

entry detail
(view a)

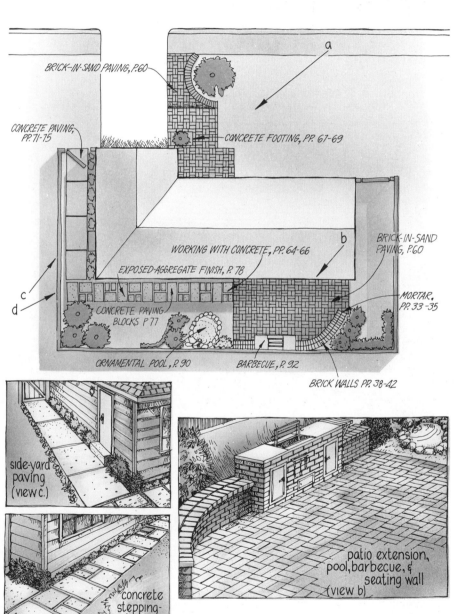

BRICK-IN-SAND PAVING, P.60

CONCRETE PAVING, PP. 71-75

a

CONCRETE FOOTING, PP. 67-69

WORKING WITH CONCRETE, PP. 64-66

EXPOSED-AGGREGATE FINISH, P. 78

b

BRICK-IN-SAND PAVING, P.60

c

d

CONCRETE PAVING BLOCKS P. 77

MORTAR, PP. 33-35

ORNAMENTAL POOL, P. 90

BARBECUE, P. 92

BRICK WALLS PP. 38-42

side-yard paving (view c.)

concrete stepping-stones (view d)

patio extension, pool, barbecue, & seating wall (view b)

INDEX

Boldface numbers refer to color photographs

PHOTOGRAPHERS

Richard Fish: 10 bottom. **Scott Fitzgerrell:** 1, 5 top right & bottom left, 13 bottom right, 19, 32. **Dorothy Krell:** 13 top right. **Steve W. Marley:** 5 middle left, 25. **Ells Marugg:** 3, 4 bottom left, 8 left, 11 top, 12 bottom, 15 top right, 22 top, 23 bottom. **Jack McDowell:** 2, 4 top right & bottom right, 5 bottom right, 6, 7, 8 right, 9 bottom, 10 top left, 11 bottom left & right, 12 top left & right, 13 top left & bottom left, 14, 15 bottom right, 17 top right, bottom left & right, 18, 20 top left & right, 21, 22 bottom, 23 top, 24, 27 top left & right, 28, 29 bottom, 30, 31. **Bill Ross:** 16 top right. **Rob Super:** 4 top left & middle right, 5 top left & middle right, 9 top, 10 top right, 15 left, 16 top left & bottom, 17 top left, 20 bottom, 27 bottom, 29 top.